I0490431

THE IMMORTAL SCIENCE

WHO PLANNED THE UNIVERSE?

VOLUME 1

VIJAY SINGH RAJPUT

INDIA · SINGAPORE · MALAYSIA

Notion Press

No.8, 3rd Cross Street, CIT Colony,
Mylapore, Chennai,
Tamil Nadu – 600004

First Published by Notion Press 2020
Copyright © Vijay Singh Rajput 2020
All Rights Reserved.

ISBN 978-1-64850-710-6

Contents

Preface

This book is based on the scientific study of Deities and the universe is not beyond the logics, GOD himself is the super scientist and his invention in the panorama of mankind are great which cannot be listed in numerous books. Peoples believe that god is beyond the logic this book on the research and verified by their scientific instrumentation, which will prove the god is logical.

The time scale of the universe is a fascinating theory in the Universe. One century ago, it was thought that the world is dominated by the order and action of Vedic god, the Trinity. I described what Vedas says, from where the vast space came from and how celestial bodies are hierarchically embedded in space.

This believes has radically changed in our days, by day to day observation. Many observational data, enriched by theoretical results, reveal the same concept of origin of the universe which was endorsed by Vedas and Puranas.

This scientific epic is based entirely on the scientific observations in Vedic view; I tried to decode the secrets of Vedic Theories, by my many years of research. The readers will enjoy whole creation in a chain tale from **Singularity** to the formation vast universe, where the celestial bodies are planned and arranged to expedite their distinct role to run the cosmic order by Shiv and Adi -Parashakti.

The entire narration is the theory of scientific phenomena by tenets of hymns of Vedas and Puranas, which shall be enjoyed by readers in the form of Puranic tales.

The book is based on the chronology of universe in Vedic scientific way, in which Mass, Energy Space and Time are the result of Anahad Naad which is called as Big Bang by modern science.

The Vedic science not only narrates the cause of Big Bang but also the state before the Big Bang i.e. Anahad Naad. Phase of inflation in the form of Maha Vishnu and the beginning of Time as a Brahma, these are the phenomenon achieved by Mass and Energy reaction.

The summon of Rudra's and the onset of war between matter and antimatter, the grand unification theory (Ishaan Shiva) and splitting of four fundamental forces in the form of four Rudras.

The story starts with the appearance of Pashupat the radient effulgence in the form of **cosmic microwave background (CMB)** is an electromagnetic radiation.

Then the beginning of various epochs at cosmological time scale starts as:

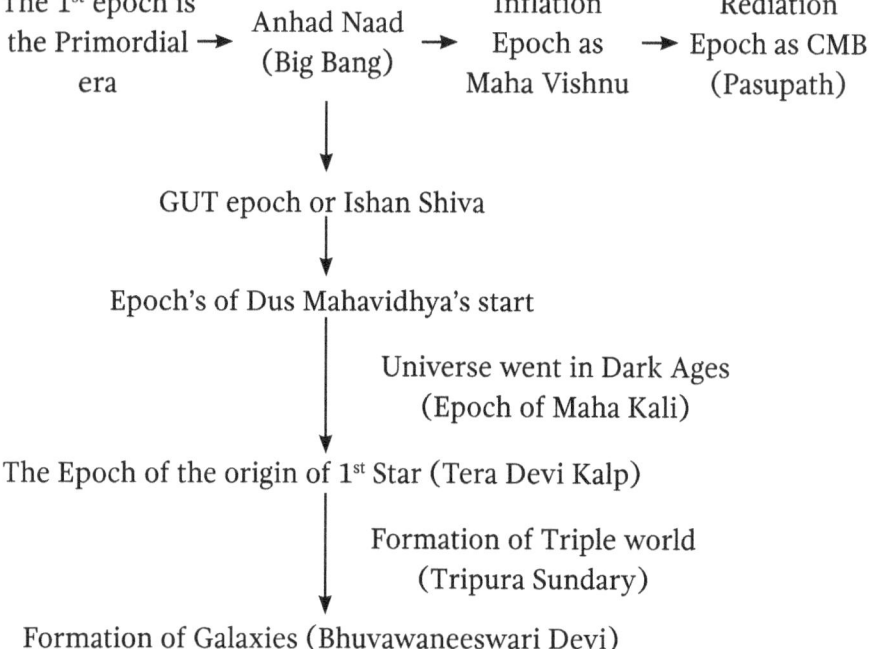

Origin of 1ˢᵗ Star of cosmos in the form of the Tara Devi, thereafter energies transform in the aspect of Dus Maha Vidhyas one after another epochs, through the chronology of origin of the universe, which had given structure to the cosmos. The burst of 1ˢᵗ Star in order to generate the new generation population III star's and are the cause and effect of Dus Mahavidiyas.

The human being at present era believes that it's the ideas of their mind that creates the new inventions which change and improve the world. We are so full of pride for our knowledge, that we have forgotten that the tools we use are the inventions of God.

This book will radically change the concept of thoughts and no matter what religious background we come from, in fact we will perceive that almighty is in every aspect of universe and every slant of the cosmos is the god.

Chapter 1

The Myth of the Beginning of Time

Jai   Shri   Ganesh

Om:- Let's begin with uttering OM, the 1st sacred sound and the ultimate reality of Universe.

Vijay:- Recently, I have gone through the article of Scientific American's and the topic was:

Was the big bang really the beginning of time? Or did the universe exist before then? Such a question seemed almost blasphemous only a decade ago.

Albert Einstein, in his theory of special relativity, determined the laws of physics that instigated new foundation for the Astronomers and physicists of our generation and offered new concepts of space and time. The theory holds that space and time are soft, malleable entities. On the largest scales, space is naturally dynamic, expanding or contracting over time, carrying matter like driftwood on the tide. Astronomers confirmed in the 1920s that our universe is currently expanding: distant galaxies move apart from one another. One consequence, as physicists Stephen W. Hawking and Roger Penrose proved in the 1960s, that time cannot extend back indefinitely. As you play cosmic history backward in time, the galaxies all come together to a single infinitesimal point, known as a singularity--almost as if they were descending into a black hole. Each galaxy or its precursor is squeezed down to zero size. Quantities such as density, temperature and space-time curvature become infinite. The singularity is the ultimate cataclysm, beyond which our cosmic ancestry cannot extend.

So, when did time begin? Modern science and astronomy does not have a conclusive answer yet.

Shital:- What does you mean by this, is there any authenticated testimonial theory which can be endorsed as a master theory, the whole time theory to encompass the cosmic enigma, the theory upon which all modern physicists should relay and to bring all on the common node?

Vijay:- Shital, I shall narrate you and clear all the doubts mentioned in the article in very blissful way, were all the scientific theories presented as of now by modern science, will conjointly come to one end and that too I will do with the help of **Vedic scientific** theories.

My narration should not be taken as a token of religion but it's a scientific conviction. I mean, it is not about the particular religion, it's all about the scientific belief, scilicet religion may change from person to person, even the science may change with the new researches and discoveries but Vedic Science will always remain as one.

This is the journey of universe from the single point (Bindu), which is as called Nothingness for the human intellect, It cannot be imagine even by Deities, no supper human has ability to detect the bindu. Where that point (Bindu) is there in this vast universe. It's like searching the lost needle in the ocean.

Shital:- Then how it can be claimed that the universe begin from Bindu?

Vijay:- There are numerous references in Vedic and Puranic cult, and same was accepted by modern astronomers like Albert Einstein's and many more. Let's see some references:

In 1935 Einstein Prodolsky and Rosen challenged Quantum Mechanics on the grounds that it was an incomplete formulation. They were the first authors to recognize that quantum mechanics is inherently non-local, which means it allows for instantaneous action across arbitrarily great distances. So an action in one place can instantly influence something on the other side of the universe in no time at all. This very powerful paper (The EPR paper) explaining Quantum

Entanglement changed the world and alerted us to the magical implications of quantum mechanics' metaphysical implications.

In a letter of Einstein, which he wrote to Max Born, 3 March 1947, "Es gibt keine spukhafte Fernwirkung" which translates to "There is no spooky action at a distance." He did not believe in magic. He believed in science and would regularly read the Bhagavad-Gita. Einstein's famous quote on the Bhagavad-Gita is: "When I read the Bhagavad-Gita and reflect about how God created this universe everything else seems so superfluous." He also wrote in his book *The World as I See It*, "I maintain that the cosmic religious feeling is the strongest and noblest motive for scientific research".

Henry David Thoreau, American Thinker & Author: whenever I have read any part of the Vedas, I have felt that some unearthly and unknown light illuminated me. In the great teaching of the Vedas, there is no touch of sectarianism. It is of all ages, climbs, and nationalities and is the royal road for the attainment of the Great Knowledge. When I read it, I feel that I am under the spangled heavens of a summer night."

The famous Danish physicist and Nobel Prize winner, Laureate Niles Bohr (1885-1962) (pictured above), was a follower of the Vedas. He said, "I go into the Upanishads to ask questions." Both Bohr and Schrödinger, the founders of quantum physics, were avid readers of the Vedic texts and observed that their experiments in quantum physics were consistent with what they had read in the Vedas.

Niles Bohr got the ball rolling around 1900 by explaining why atoms emit and absorb electromagnetic radiations only at certain frequencies.

Vijay:- These are what modern scientists of different ages observed and believed in there findings of practical science.

I will narrate you the science of my research and observations, it's not a comparison of modern and Vedic or Puranic science, honestly it's the science coded in every epic of our heritage.

Let's begin with Nothingness and understand what exactly Nasadiya Sukta is.

1.1 Origin of the Universe (An Introduction) and Nasadiya Sukta as Theory of Nothingness

Before the origin of the universe there is only a vacuum or the Sunya (the absolute zero), the lord Shiv Shakti is one as a Singularity, particles particularly in the form of energies. Then these energies get started to transform in deferent forms as a result of which universe were created. The big bang theory is an effort to explain what happened during and after that moment.

What modern cosmology informs us about the birth (origin) of the universe today is already proven in Hindu Rig Veda long before from ancient time. The universe came into being all of a sudden, 15,000 million years ago. It was not originated with all the stars, planets and galaxies present now. It made its presence all of a sudden from nowhere with unimaginably high density and temperature, in the size of a pin tip. It contained all the matter and energy of today's universe in the form of total Mass. It has been mentioned in Rig Veda that universe was originated from the golden womb, again this golden womb (Hiranaya Garbha referred to the singularity).

A group of scientists believes the universe was originated from nothing, for which they called the theory of nothingness.

Let me state that, the theory of nothing is just nothing, while it seems to be nothing instead it is something that posses everything required to generate a universe. **Everything must have a beginning. The only logical assumption to make is that in the beginning there was nothing. It was the absolute absence of space, time and the universe as we see today. But that nothing was having something i.e. mass and energy are composed of all the universal characteristics to produce Big Bang.**

Absolute nothingness to the extent means that even nothing does not exist. The beginning starts from nothing, Due to the weirdness of quantum mechanics, nothing transforms into something at that point. **Heisenberg's** uncertainty principle states that a system can never have precisely zero energy and since energy and mass are equivalent, pairs of particles can form spontaneously as long as they annihilate one another very quickly.

Detectability is paramount in understanding logics such as Mass, Energy, Space and Time. The same is applicable to detect the existence of matter. Human perception is based on detecting the things and scientific measurement is an extension of human perception. If something is not detectable, it does not mean it doesn't exist.

Nasadiya Sukta of Rig Veda is the theory of origin of the universe which states the same what quantum mechanics says that **Absolute nothingnessis is.entirely or absolutely absence of everything.** (Sukta is the Sanskrit word, its meaning in Hindi is sutra and in English, it means a formula) It is concerned with cosmology and the origin of the universe. Here is the Sukta and its translation:

1.2 Scientific Explanation of Nasadiya Sukta

Shital:- So, you mean **Nasadiya Sukta** states the theory of the Universe origin.

Vijay:- Yes, It's better to say that Nasadiya Sukta is the theory of before and after the beginning of universe, growth and sustainability of the Universe.

Nasadiya Sukta is the narration about universe in the aspect of nothingness such as Dark Energy, Mass-Energy, Space and Time.

It is applicable to the various epochs or the timeline of the Universe. Listen to it, but when we go deep in the discussion you will understand it better.

नासदासींनोसदासीत्तदानीं नासीद्रजो नो व्योमापरो यत् ।
किमावरीव: कुहकस्यशर्मन्नभः किमासीद्गहनं गभीरम् ॥ १ ॥

nasadasinno sadasīttadānīm nāsīdrajo no vyoma paro yat |
kimāvarīvaḥ kuha kasya śarmannambhaḥ kimāsīdgahanam gabhīram ||1||

Scientific explanation:- Nothing exists before the origin of the universe.

Also, no space and no were the sky. What and where was hidden who knows that?

Neither water nor surface was there at that moment before the big bang.

As there was no Space and Time, Thus there is no way of detecting the anecdote or event before the big bang.

न मृत्युरासीदमृतं न तर्हि न रात्र्या।आन्ह।आसीत् प्रकेत: ।
आनीदवातं स्वधया तदेकं तस्माद्धान्यन्नपर: किंचनास ॥२॥

na mṛtyurāsīdamṛtaṁ na tarhi na rātryā ahnȧ āasītpraketaḥ |
ānȧdavātaṁ svadhayā tadekaṁ tasmāddhānyanna paraḥ kiñcanāsȧ ||2||

There was neither death nor immortality then,

There was no sign of night or day.

That one breathed all by self without any outside support,

Other than that there was nothing else beyond.

Scientific explanation: Without the absence of space (Vishnu) there is no option of life or death. As a whole the Mass (Shiv) is conserved in Dark Energy (Adi Shakti, Durga) there was no light. Everything was in one (Akaari) called Singularity.

तम।आअसीत्तमसा गूह्ळमग्रे प्रकेतं सलिलं सर्वमा।इदम् ।
तुच्छेनाभ्वपिहितं यदासीत्तपसस्तन्महिना जायतैकम् ॥३॥

tamȧ āasȧttamȧsā gūḻhamagré'praketaṁ salilaṁ sarvȧmā idaṁ |
tucchyenābhvapihitaṁ yadāsȧttapȧsȧstanmȧhinā jȧyataikȧṁ

There was darkness all wrapped around by darkness,

and all was water then.

The life force which was covered with void arose through

the power of heat (Tapas).

Scientific explanation: The darkness all around was the Dark Energy, Which is said to be goddess **Maha Kali,** She wrapped all the cosmic stuff in compressed form. The life or the beginnings of the Universe will occur with the heat (Tapas, means the Mass and energy, Shiv and Shakti or the character of Purusha and Prakriti within singularity)

कामस्तदग्रे समवर्तताधि मनसो रेत: प्रथमं यदासीत् ।
सतोबन्धुमसति निरविन्दन्हृदि प्रतीष्या कवयो मनीषा ॥४॥

kāmastadagre samāvartatādhi manāso retāḥ prathamam yadāsīt |
sato bandhumasāti nirāvindan hṛdi pratīṣyā kavayo manīṣā

In the beginning, Love arose,

which was primal germ cell of mind.

The Seers, searching in their hearts with wisdom,

discovered the connection of being in Nonbeing.

Scientific explanation: There was that Parbrahm or Shiv, the mass less (Aghan Awastha) and it was his will and desires to reproduce the universe. This was known to be founded in the heart of ancient researcher's or Sages (Scientists or Seers) about the discovery of genesis from nothingness.

तिरश्चीनो विततो रश्मीरेषामध: स्विदासी ३ दुपरिस्विदासीत् ।
रेतोधा।आसन्महिमान् ।आसन्त्स्वधा ।आवस्तात् प्रयति: परस्तात् ॥५॥

tiraścīno vitāto raśmireṣāmadhaḥ svidāsī 3 dupari svidāsī 3 t |
retodhā āasanmahimāna āasantsvadhā āvastātprayātiḥ parastāt

A ray of light energy cut across the dark and gloomy abyss.

Was it beneath? Was it above? Who can answer this?

There were bearers of seed and mighty forces,

Pushed from below and forward move above.

Scientific Explanation: Step by step, State the beginning of the Universe is mention in this stanza.

i. A ray of light means, the energy generated from Mass within Singularity (Sakal Shiv and Uma Devi).

ii. तिरश्चीनो विततो रश्मीरेषामध: स्विदासी means: Mass and energy reaction resulted inflation of space as Maha Vishnu.

iii. This was the point that can be detected, hence cosmic Time as Brahma begins here.

iv. रेतोधा।आसन्महिमान् ।आसन्त्स्वधा: The four unified fundamental forces generated and became separate forces of nature.

Universe With the appearance of Shakal Shiva (Adi Shakti and Lord Shiv in Physical form) now here is the stage of origin of big-bang was the begotten of Lord Maha Vishnu with the inflation of space. All the fundamental forces are united and were separated out.

को अद्धा वेद क इह प्र वोचत्कुत आजाता कुत इयं विसृष्टि:।

अर्वाग्देवा अस्य विसर्जनेनाथा को वेद यत आबभूव॥६॥

ko addhā véda ka iha pra vócatkuta āajātā kutá iyaṁ visṛṣṭiḥ |
arvāgdevā asya visarjánenāthā ko véda yatá āababhūvá ||6

Who knows? Who can confidently declare it?

From which was it born? Who gave rise to this creation?

Even the Gods came subsequent to creation,

then who can reveal from where it arose?

इयं विसृष्टिर्यत आबभूव यदि वा दधे यदि वा न।

'यो अस्याध्यक्ष: परमे व्योमन्त्सो अङ्ग वेद यदि वा न वेद॥ ७॥

iyaṁ visṛṣṭiryatá āababhūva yadi vā dadhe yadi vā na |
yo asyādhyákṣaḥ paramé vyómantso aṅga véda yadi vā na vedá || 7 ||

That out of which creation arose,

whether it formed by itself or it did not,

He who oversees it from the highest heaven,

only he knows or maybe he does not.

Shital:- Is Nasadiya Sukta is a complete theory of origin of the Universe?

Vijay:- To know the exact and complete meaning of the last stanza which put forth a question, who knows it?

I will reveal this phase by phase and explain the chronology of the Universe of all epoches.

It is simply understood that, the scientist from all around the world is working and trying to resolve the quarries about the universe. That from where it was originated, how it works and will it end?

So the simple answer to your question is that. Almighty blessed us recently a Daughter. We call her **Shivomya.** Now let you answer me that, what do you know about her and where she was earlier?

Shital:- Oh she is just a month old, I don't know much but just I was known by her movements when she was in my womb and before that where she was I don't know.

Vijay:- Ok so take her now as a universe, but you don't know before the 9 months of her in your womb. So that universe (Shivomya) was not present in your experience that means there was nothing that you think.

Same are the doubts of some modern scientists around the world some are busy in proving the early theory as wrong while many are doing thire work of new discoveries with full dedication.

I am taking a few statements of some scholar who are trying to prove Big-Bang theory wrong.

I think, this is the estimation of doubts knocking the doors of minds these scientists. Now time came to clear their doubts through the knowledge of Vedas and Puranas.

* * *

The words of some scholars

"The Big Bang is pure presumption. There are no physical principles from which it can be deduced that all of the matter in the universe would ever gather together in one location, or from which it can be deduced that an explosion would occur if the theoretical aggregation did took placed.

"Theorists have great difficulty in constructing any self-consistent account of the conditions existing at the time of the hypothetical Big Bang. Attempts at mathematical treatment usually lead to the concentration of the entire mass of the universe at a point.

"'The central thesis of Big Bang cosmology,' says **Joseph Silk** *'is that about 20 billion years ago, any two points in the observable universe were arbitrarily close together. The density of matter at this moment was infinite.'*

"This concept of infinite density is not scientific. It is an idea from the realm of the supernatural, as most scientists realize when they meet infinities in other physical contexts. **Richard Feynman** *puts it in this manner:*

"'If we get infinity [when we calculate] how can we ever say that this agrees with nature?' This point alone is enough to invalidate the Big Bang theory in all its various forms." —*Dewey B. Larson, the Universe of Motion (1984), p. 415.

"In fact, evolution became in a sense a scientific religion; almost all scientists have accepted it and many are prepared to 'bend' their observations to fit in with it." —*H. Lipson, "A Physicist Looks at Evolution," Physics Bulletin 31 (1980), p. 138.

They are having doubt in nothingness which means it is nothing because it never comes in proof, just because you can't see **it doesn't mean it doesn't exist.**

This seems to mean there must be some evidence that the object exits. However, the object I'm looking for does not have any evidence that it might exist.

Practically, it is possible that it might exist but human errors are always possible, so which never has their traits is called nothingness.

But that nothingness is truly the **Singularity** which is even mystery from Einstein to modern Scientists but all strongly believes in it.

Now let me break the theory of nothingness, which says that the universe came from nothingness is only a partial theory, like you knows about our daughter's presence and origin.

Our daughter got origin from something called genetic character which was in both of us as a matter (mass). The pure soul as Space is the embodiment of eternal bliss comes from Lord Narayan (Vishnu). Then the Energy (Shakti or Adi Shakti) evolve. By the combination of three i.e. mass, energy and space **SHIVOMYA** appeared, that moment was called Time. Thus nothingness itself came from something having all the characteristics required to form a universe.

1.3 The Singularity and Its Explanation through the Formula of Kunjikastrotram

According to the standard theory of Relativity, our universe sprang into existence as "singularity" around 13.7 billion years ago. Singularities are zones that defy our current understanding of physics. They are thought to exist at the core of "black holes." Black holes are areas of intense gravitational pressure. The pressure is thought to be so intense that finite matter is actually squished into infinite density (a mathematical concept which truly boggles the mind). These zones of infinite density are called "singularities." Our universe is thought to have begun as an infinitesimally small, infinitely hot, infinitely dense, something - a singularity. The Singularity was Lord Shiv Shakti the Eternally Limitless Power". That is, it is the Power before this universe. It is the active energy that creates and dissolves the entire universe. Some sacred texts state that she is the goddess **Bhuvaneshwari** or Adi-shakti.

A black hole is created when matter (Maha Rudra) and space (Narayan) become concentrated so that there is no differentiation between them. The black hole is only detectable by its effect on the surrounding matter and space. Space i.e. Narayan and Rudra i.e. matter with the effect of energy (Adi Shakti) at unimaginable velocities approaches the event horizon.

This is the reason why both are respecting each other and both are equally worshiped. The whole universe exists because of energy, even it has been originated because of energy and that it is the capacity of a system to perform work, the simplest is that, it is an energy density inherent to empty space and that is the dark energy.

In physical cosmology and astronomy, **Dark Energy** is a hypothetical form of energy that permeates all of space and tends to accelerate the expansion of the universe. Dark energy refers to be Adi Shakti the energy of the whole universe. This is the energy which is keeping the planets in balance, galaxies in motion and interconnected all celestial bodies with each other.

The Science of Kunjikastrotram (Key formula which unlocks) has been told to goddess Parvati by Shiv (Reference- Durga Saptasati). This is also the truth about the origin of the Universe. Kunjikastrotram is the decoded form about the birth of the universe to the sustainability of livingness.

Once Goddess Parvati asked to Shiv

Devi Parvati asked:- O Lord, What is the secrets code of the whole universes?

Shiv Says: Devi, listen to the secrets of secret code how the universe begin and sustain.

He Said **Om *Aim Hrim Klim Chamundaye* Vichhe Namaha.** This is the code by which the Universe got origin, by which life begins and by which desires sprouted in the brain.

Aimkarisrushtiroopayaihreemkariprathipalika, **Kleemkari Kama Roopinyai, bheejaroopenamosthuthe,**

Chamunda chandagathi cha yeikari Varadhayini, Viche chaabha yadhanithyam namasthe Manthraroopini.

Hey Devi the Adi Shakti in the form of **Aimkarisrushtiroopayai** or Akari i.e. She only (alone) one Divine Dark energy was the supreme energy exists in the form of Singularity (Aimkari) have created (Srusthi) and shaped (Roopayai) the whole universe.

Aim is the Adi-Shakti or Uma Devi from her, the universe got begotten.

In the form of **Hreemkari prathipalika** or Maha Lakshmi is nourishing (prathipalika) the universe.

Kleemkari Kama Roopinyai, bheejaroopena namosthuthe, or in Klim (Saraswati) form is giving the desire and sowing the seeds to reproduce lives.

And **Chamundey** is the destroyer of all evils and demons (chand or chandals).

Vichay is the remover of all fears and boon givers.

We will discuss in more detail about the scientific phenomena of Kunjikastrotram in coming episodes in the timeline of the universe.

Chapter 2

Shiv the Mass, Adi-Shakti the Primordial Energy, Vishnu the Space and Brahma the Time

Shital:- I want to know the secret of whole science endorsed in Vedas, to oblige on those who wish to know more about the scientific phenomena and actual evidence of the origin of creation. As we see modern theories, there are so many conflicts that do not support each other.

Some Scientist proved and believed in the theory of relativity, some believes in quantum mechanics while some believes in the string theory, also there is a community who trust in the theory of evolution.

Many theories are paradoxical and have contradictory statements. On which scientific notionc can universe stand on a common node of belief and trust?

Vijay:- Entire observable and even unobservable universe is the record of evidences and these are embedded in ancient Vedas. So the one and only one evidently and trusted science is Vedic science, all sub-theories are put forth, from the root of ancient Vedas.

"Our universe may be viewed in many ways- by mystics, theologians, philosophers and scientists. In science we adopt the practical route: we accept only what is tested by experimentation or observation."

Same way Vedas and Puranas are tested by ancient Scientists (Rishis). What they experimented and observed in their whole life, which is dedicated to realize supreme and his creation through their research, they have coded in Vedas, Puranas and in Upnishads.

The Seer Vidac Vyas around 5000 years ago done the job of simplifying the four Vedas and categories them into eighteen Maha-Puranas to understand the Sanskrit text better for forthcoming generation, Scientists or the Seers of different ages of this Kali-Yuga.

But the real science of these Vedas, Puranas and Upnishads are still mysterious for the modern Scientist, then how it can be known by common peoples.

Vedic Science is hard to define and it's even far beyond the Ancient Anthropology, the **Vedas** are the source of integral wisdom, science, tradition which states, Adi-Parashakti: The empyrean Goddess, Devi is the Supreme Being and recognized divine, the superior mother. I strongly believe that the theory of divine Energy is the base of all theories of energy.In modern era, scientific and engineering developments involve energy concepts from many scientific disciplines. The Vedic formulas considered to be sacred by the Vedic religion states.

The Devi Bhagwata Mahapurana states that Adi-Parashakti, the Dark energy is the original creator, observer and destroyer of the whole universe. Hence AdiParashakti is Param Prakriti or Supreme Nature, the Parvati, the goddess of power is considered as her Sagun Swaroop (human form). That is to say that Parvati is the Goddess, comprised of the three Gunas, or qualities: Sattva, Rajas, and Tamas. Sattva is balance or equilibrium; Rajas is restlessness or imbalance; Tamas is inertia or darkness.

However, In the beginning of the universe before the first Kalp (Kalp means eon), the divine dark energy which exists as a singularity get splits in the Mass (Sagun Shiva) *by its own will into two equal nuclei of negatively charged ions and positively charged ions having a huge amount of mass and energies which are Shiv and Shakti by a binary fission processes.*

Fission is a form of nuclear transmutation because the resulting fragments are not the same element as the original atom. The two nuclei produced are Shiv Jyothilinga and Adi-Shakti, from these Maha Maya by ternary fission produced three supreme forms of feminine energies as Maha

Lakshmi, Maha Kali and Maha Saraswati most often of comparable but slightly different.

"The three Guans (qualities) are Rajgun-Brahma Ji, Satgun-Vishnu Ji, and Tamgun Shiv Ji. They have taken birth from Brahm or Shiv (Maha Kaal) and Prakriti (Durga) and all three are perishable."

Evidence:- Shri Shiv Mahapuran

Shital:- What is Brahm, Brahma and Brahmand?

Vijay:- Shital, you have raised a very interesting question: there is a distinction between things that are alive and things that are not or we can say there is a continuum of things that cover the spectrum from living to non-living. To the best of my current knowledge, the distinction between the three sets is reasonably sharp. However, you made me curious. I'd like to narrate your this question in more detail. So Listen:

The Brahm is the combined super divine soul of the three qualities mass, space and time and Prakriti the Dark energy itself is the nature which covers all these three things before and after the Big Bang within her in the form of Singularity.

Brahm is what always exists from which universe (Brahmand) get originate and Brahma is the time come in existence just after Mass (SHIV) and Space (VISHNU) immediately after Big bang. Therefore Brahma is the discoverer of the whole cosmos.

But before knowing and understanding Brahm we need to have all our devotions in Sada-Shiv, Adi-Shakti, Maha-Vishnu and Brahma. I mean to say that we should be very clear with Mass, Energy, Space and Time.

Second Vedic evidence:- *Shrimad'devibhagwat Puran,* - God Vishnu prayed to Durga: said that I Vishnu (Space), Brahma (Time), and Shankar (Matter) are exists by your grace. **We have birth (aavirbhaav) and death (tirobhaav).** We are not eternal (immortal). Only you are eternal, are the mother of the world (Jagat Janani), are Prakriti, and Goddess Sanatani (existing for time immemorial).

Shankar said: If Shri Brahma and Shri Vishnu have taken birth from you, then am I, Shankar (Rudra), who was born after them and perform Tamoguni Leela (divine play), not your son? Henceforth, you are my mother too. Your gunas are always present everywhere in this world's creation, preservation, and destruction. The three of us, Brahma, Vishnu, and Shankar, born of these three Gunas (qualities) remain devoted to work according to the regulations.

Kalika Purana says, Tino Devon Ka Anantavya:- "When all the Bhuvanas/Lokas/worlds were enveloped with the darkness of the deep, they were not visible. "There was neither the division of the day and night nor were the five elements (Panch Maha-Bhutas as Earth, water, fire, air and sky). Parabrahma represents His super Grand Unified field 'Om' surpassing the Einstein's concept because it includes all knowledge as well. He was the only one in minutest form; He is immortal and indescribable in dualistic form from the knowledge point of view. He/She is the only singularity "Both the "Prakriti" (nature) and Purusha are perpetual. Time is seated there, which is the only cause of the universe. "He is formless and for that particular universe, He is present always in three forms. "Time has a form- a pattern - a cause that is associated with the Bhutas. "He reveals Himself by his radiance. From the very beginning, He is possessed because of His keen desire. "When the Prakriti is distorted and perturbed, Mahatatwa is born and thereafter three kinds of ego prevail. "With the ego, sound Tanamatra was born and Vishnu created an infinite sky having no image.

Meanwhile, by the kinetic source nuclei, Shiv Jyothi Linga stimulated these three feminine super energies due to which Maha Lakshmi Split into isotopes Brahma and Lakshmi, from Maha Kali the Shankar and Saraswathi, and Mahasaraswathi produced Shiva and Vishnu.

Vijay:- These are the highlights to know what happened in the beginning of the universe, what causes to bring the present universe in such a big way. According to big bang theory, fifteen billion years ago, our universe was compressed into a point of the size of an embryo. Known as a singularity, this is the moment before creation when space and time did not exist.

Before the beginning there was nothing, said by many viewers and scientists but it was proven that there was something called singularities. In which everything exists in a compressed form. Just after the big bang explosion, the universe born and that was the beginning of mass, energy, space, and time.

Then Brahma companion with Saraswathi and started the cosmic life creation. By doing so Brahma Ji got the designation of **Prajapati** (in Sanskrit Praja means the Folk or People or inhabitants) as he is the creator of whole creation hence he is the Prajapati of the entire creation.

Vishnu with Lakshmi started to feed and nourish all lives. Shankar along with Shiva (Devi Uma) started to destroy all the unwanted particles. Therefore this is the common belief system of Sanathan Dharma. Philosophically and fundamentally this is true sect.

The Mass is Sada-Shiv, energy is Adi-Shakti, Space is Maha Vishnu and time is Brahma. Shiv is always there before the beginning and after the dissolution of the universe in the form of **Mass**. The lord Maha Vishnu as **Space** is having his existence in entire of the universe. So mass and energy are there with in the singularity. Also, Space as Maha-Vishnu has always his existence therefore cosmic space is the vacuum that exists between the celestial bodies. Thus Maha Vishnu is regarded as a Parmathma or super soul, Pratham Purush or the first man. Therefore mass, energy and space is directly proportional to each other and are not different from one another.

Therefore after the big bang (Anhad Naad) it experienced an incredible burst of expansion known as inflation, in which space (Vishnu) itself expanded faster than the speed of light.

Thus the **detectability** of Mass and Energy in shapeless (Nirgun) is Jyotilingum and symbolised shape is Ardnarishwar Swaroop.

In other words, Shiva the Supreme Being is outside as well as inside of space, time and matter. He is not constrained by the limits of his creation, nor by the wonderful laws that govern it; of gravity, of the speed of light, of the charge of an electron. The universe was not originated with all the stars, planets and satellites present now. It made its presence all of a sudden from nowhere with unimaginably high density and temperature, in the size of a pin tip. The pin is said in Rig Veda a BINDU i.e. the Brahm. Bindu is the point in which whole energy, mass and even space of today's universe were present. **BRAHM (ब्रह्म)** is small as point and big that as of whole universes.

2.1 Brahma the Time

Vijay:- Shital listen, to understand the absolute cosmology in Vedic and modern scientific perspective, you need to know first the Mass, Energy, Space and Time. I would like to narrate you the scientific phenomena of the beginning of Time in the reference of Brahma as an Individual character.

Cognizant of this amazing event that, Time as a Brahma doesn't mean the periodic succession of day and night or the position of celestial bodies in the sky. Time in cosmology is not something like the appearance of the sun on dawn in the horizon.

Brahma Ji cannot be characterized something like the position of the sun in the sky or the marking the moment of the morning, noon and evening time during the day.

Brahma Ji is the one the primordial being, evolved by the expansion of space and with his beginning, the real-time started.

The idea of cosmic contraction/expansion linked to gravity and entropy yields the special definition of time: Time is the measure of the universal progression of uniformity between matter and space, accomplished by counting equal, standardized divisions of a cyclical system of regular motion.

The Time by scientific means is the Brahma which began immediately after the big bang. That's why the Vedic mythology stated the time scale as a Kalpa or the eon. We can also say that the Cosmic Time began as the birth of Brahma and this universal Time will remain up to the final **cosmic cataclysm** and that will be the end of Time, at that moment again everything merges in singularity.

The embryonic universe was still expanding fast. Now that's the reason why nobody knows beyond time because even an imaginary line cannot be drawn beyond the time. So there is no proof of the universe and its existence without Brahma. By the origin of Brahma time (KAAL) was discovered.

Now the question is how the Brahma or the Time comes in its active state? Brahma gets sprouted from the navel of Lord Maha Vishnu and gets attached through the umbilical cord with the belly button. The umbilical cord that joins mother to child, nourishing and cleansing, supplying the growing fetus with its every need, until the time of birth, when mother and infant finally meet face to face i.e. Brahma and Vishnu are face to face. The navel of Vishnu is the kundalini Shakti or the feminine energy of Adi Shakti or the Yog Maya from which infant Brahma is getting the nourishment. The lotus over which Brahma is seated was like the amniotic sac protecting him from outer infection.

The Inflation of Space as Maha-Vishnu and the origin of Time as Brahma proceeded by the expansion of Matter in the form of Galaxies

By the birth of Brahma, the universe got originated and that was an actual period of detectability. If there was no time there will not any process of progression.

One microsecond (Laghu - Chhan) after the Big Bang, all of a sudden all the fundamental forces involved in the structure of matter came into being such as the 'strong', 'weak' and electromagnetic forces. The most basic building blocks of matter, namely the quarks were created. Quarks joined together giving rise to the fundamental particles the electrons, protons and neutrons. All this happened step by step in several sub-stages.

By the hundredth second, the nucleus of helium was constituted, and it took 100,000 years for a few nuclei such as those of hydrogen to form.

300,000 years elapsed before electrons got attached to nuclei so as to form atoms of these few elements. The embryonic universe was still expanding fast. The matter created was a vast body of dense smoke, a cloud of dust. It was dark. By the end of this period, it became more transparent and bright with light.

2.2 Inflation or the Expansion of Lord Maha Vishnu as Space

Vijay:- It is believed by modern science that the very early rate of expansion of the universe was much faster than today, it could not have been had the gravity. This period of rapid expansion is referred to as inflation. It is thought that inflation occurred particularly because of the physical force of gravity, as we know it did not yet exist (temperatures were too high, distances too short). There was a repeated collapse between the stars. Then the Brahma started to create life and initially, he has created sun god from the image of Narayan and known by the name of Surya-Narayan. He is the representative of light and attracts all the helium and most of the hydrogen atom from that universe thus the most of the cosmic universe is now free from highly helium plasma.

That momentum and subsequent separation of matter into a huge, expanding cloud, is taken to be the fundamental origin of all energy, subsequently generated by gravitational collapse, following this brief period of inflation, the initial, extremely hot universe began cooling right away (expanding gasses cool). An important consequence of cooling of the universe is that subatomic particles were able to condense into the matter that we recognize in today's universe. These include such things as electrons, protons and neutrons. Later, with additional cooling (about 1 million years), there was a condensation of subatomic particles into hydrogen atoms. In addition to the creation of these condensed forms of matter, as temperatures declined, there was a similar formation of the physical forces recognized as operating in the universe today.

2.3 Space the Aspect of Maha Vishnu

Vijay:- The entire of the space is Maha-Vishnu, whatever that exists apart from Mass (in the form of Matter) throughout the universe is Space referred to as Vishnu or Narayan in Vedas.

Mass is Sada Shiv as all the substances or the material of cosmos even the whole universe is the part of Shiv.

Lord **Narayana** is described as having the divine black and blue color of water-filled clouds, four-armed, holding a Padma (lotus flower), mace Kaumodaki, Panchajanya Shankha (conch) and a discus weapon Sudarshana Chakra in the scared text of Vedas and Puranas. Narayana is also described in the Bhagavad Gita as having a 'Universal Form' (Vishvarupa) Vishvarupa which is the universal form of God.

The color of Lord Vishnu is cloudy blue (Megha-Verna) is the Color of the sky that means omnipresence of Vishnu.

Shital:- Why is complexion of Idols of Lord Vishnu (Rama, Krishna and his other incarnations etc.) in temples depicted as blue and black?

Vijay:- I will answer you this in Vedic view, with the reference of modern astronomy subjected as Star Child Question of the Month for December 2002

Now yesterday on 7ᵗʰ Apr'19 Shivomya is explaining about Narayan Shri Maha Vishnu to her Uncle Sanjay Singh.

Shital:- "Oh what She told him about Lord Vishnu"?

Vijay:- I *observed while sitting* down for lunch, She was narrating about the presence, attribute and features of Lord Vishnu. She told that Shri Narayan is everywhere, inside us, surrounding us and universally the vast space.

So it's like same what is referred to as Star Child Question of the Month for December 2002 posted by NASA and the question is: Why is space black?

NASA answered: Your question, which seems simple, is actually very difficult to answer!

It is a question that many scientists pondered from centurie including Johannes Kepler, Edmond Halley and German physician - astronomer Wilhelm Olbers.

There are two things to think about here. Let's take the easy one first and ask "why is the daytime sky blue here on Earth?" That is a question we can answer. The daytime sky is blue because light from the nearby Sun hits molecules in the Earth's atmosphere and scatters off in all directions. The blue color of the sky is a result of this scattering process. At night, when that part of Earth is facing away from the Sun, space looks black because there is no nearby bright source of light, like the Sun, to be scattered. If you were on the Moon, which has no atmosphere, the sky would be black both night and day. You can see this in photographs taken during the Apollo Moon landings.

So, now on to the harder part - if the universe is full of stars, why doesn't the light from all of them add up to make the whole sky bright all the time? It turns out that if the universe was infinitely large and infinitely old, then we would expect the night sky to be bright from the light of all those stars. Every direction you looked in space you would be looking at a star. Yet we know from experience that space is black! This paradox is known as Olbers' Paradox. It is a paradox because of the apparent contradiction between our expectation that the night sky be bright and our experience that it is black.

Many different explanations have been put forward to resolve Olbers' Paradox. The best solution at present is that the universe is not infinitely old; it is somewhere around 15 billion years old. That means we can only see objects as far away as the distance light can travel i.e maximum up to 15 billion years. The light from stars farther away than that has not yet had time to reach us and so can't contribute to making the sky bright.

Another reason that the sky may not be bright with the visible light of all the stars is that, when a source of light is moving away from you, the wavelength of that light is made longer (for light, means more red.) This means that the light from stars that are moving away from us will become shifted towards the red, and may shift so far that it is no longer visible at all. (Note: You hear the same effect when an ambulance passes you, and the pitch of the siren gets lower as the ambulance travels away from you, this effect is called the Doppler Effect).

Note:- The StarChild site is a service of the High Energy Astrophysics Science Archive Research Center (HEASARC), Dr. Alan Smalc (Director), within the Astrophysics Science Division (ASD) at NASA/ GSFC.

Vijay:- It's interesting, the question referred to as Star Child Question, was answered precisely in the same pattern by Shivomya as a Child and same was done by NASA.

I will elaborate the description of Narayan in Vedic Scientific perspective, this might be pious but more evidently it's a scientific phenomenon observed by ancient seers from the primordial past of their everyday life. The Shri Hari Vishnu or Narayan is regarded as Blue color when they were observed as the vast space from the earth during day time. But the same Space when they observed in the state of deep meditation or during night hours, they found the Shri Vishnu Black colour. Vedas described Black Color because Lord Vishnu has been detected as Black color by Celestial deities.

This will be cleared by following Slok:

शान्ताकारं भुजगशयनं पद्मनाभं सुरेशं
विश्वाधारं गगनसदृशं मेघवर्ण शुभाङ्गम् ।
लक्ष्मीकान्तं कमलनयनं योगिभिर्ध्यानगम्यम्
वन्दे विष्णुं भवभयहरं सर्वलोकैकनाथम् ॥

Shaanta-Aakaaram Bhujaga-Shayanam Padma-Naabham Sura-Iisham

Vishva-Aadhaaram Gagana-Sadrsham Megha-Varnna Shubha-Anggam|

Lakshmi-Kaantam Kamala-Nayanam Yogibhir-Dharyaana-Gamyam

Vande Vissnnum Bhava-Bhaya-Haram Sarva-Loka-Eka-Naatham ||

The One who rests in all and within him is everything nothing is beyond him. It is the space where the infinite universe exists.

Santakarnam- ("Santa" i.e.Silant + "Akaram" i.e. Shape) the one who has Serene Appearance is the space.

Bhujagsainam – ("Bhujag" Snakes + "Sainam" Sleeping) generally means who sleeps over the bed of Snakes. But scientifically and the real Vadic understanding is the one which is surrounded by one of the fundamental forces electromagnetism. The Sheshnag is the snake and have strong electromagnetic and Electrostatic Forces in Space associated with mass and space throughout the cosmic bodies.

The Sheshnag is the principle of electromagnetism and is the one form or the embodiment of Mahadeva Shiv Shankar, for the reason he is also called Sankarshan (Mean the one transformed from Shankar)

Padma-Naabham ("Padma" Lotus-"Naabham" Navel) who has lotus navel like universe where Brahma Ji is engaged in accomplishing the task of creation.

Suresam – Lord Vishnu is the God of Gods.

Vishvadhaaram ("Vishva" World + "Aadhaaram" Base) That Space Maha Vishnu is the sustainer of universes. Vishva (entire Matter) is dwelling in the Space.

Gagana-Sadrsham ("Gagan" Sky – "Sadrsham" Appear's)- The one who is Boundless and appears infinite like the Sky i.e. Space.

Megha-Varnna ("Megha" Cloude – "Varnam" Colour) Lord Maha Vishnu looks like the color of Bluish Clouds. That is why Vedas state's Lord Vishnu in blue appearances. And his incarnations are the cloudy colour in appearances like Shri Ram and Krishna.

When Shri Vishnu incarnates on earth, his features resemble Blue and Blackish Color in the same way as Space or Sky seen from earth atmosphere. Shri Vishnu is the Vast Space of Black color when manifested on earth scattered with the atmosphere and his morph's looks like Blue because Blue light is scattered more than other wavelengths by the gases in the atmosphere and so what Lord Narayan does on earth.

*Sunlight reaches Earth's atmosphere and is **scattered** in all directions by all the gases and particles in the air. Blue light is scattered in all directions by the tiny molecules of air in Earth's atmosphere. Blue is scattered more than other colors because it travels as shorter or smaller waves. This is why we see Lord Vishnu incarnations and his idols like Lord Shri Ram and Shri Krishna in blue and black complexion on earth.*

Shubha-Anggam ("Shubha" Good- "Anggam" organs or body parts). Who has a Beautiful and Auspicious Body.

Lakshmi –Kaantam ("Lakshmi" Goddess of wealth –Kaantam" Husband) who is the Husband of Devi Lakshmi a form of wealth energy.

Kamala-Nayanam Yogibhir-Dhyaana-Gamyam "Vande Visnum Bhava-Bhaya-Haram".- Salutations to that Vishnu, who is the remover of fear.

Sarva-Loka-Eka-Naatham - He is the Lord of all the Lokas and entire cosmos.

Space is represented by Lord Vishnu, which is omnipresent, his presence is everywhere. Etymologically and logically Vishnu is one who pervades and who entered in everything.

What modern science says bout Space is just a partial truth of their research? It has to go long miles to get the exact understanding of Space. Though what is proven till today is the same which was given in Vedas. Modern Science has a bit of understanding about reality but the Vedas and Puranas are having the ultimate truth about the cosmos. In future, whatever will be the scientific discoveries that we all will find decoded in Vedas.

The problem with this era is that only about 1% of the worldwide population is having an understanding of Sanskrit. Even those who know Sanskrit don't have that ability to unfold the truth. Nothing is hidden in Vedas, in fact we are failed to understand the deep meaning rooted in Vedas, Puranas and in Upnishads.

Science says Space is the boundless three-dimensional extent in which objects and events have relative position and direction. Physical space is often conceived in three linear dimensions, although modern physicists usually consider it, with time, to be part of a boundless four-dimensional continuum known as space-time, the same has been declared by Vedas and Seers of ancient time. The day will come when the science on its zenith will state Space in its infinite Form as Maha Vishnu as a Vishwa-Roopam. The fact is that universe is embedded in multiverse dimensions.

Modern concepts and theories of Cosmology postulate multiple universes and multiple dimensions that give the universe its structure. The structure of universe is only with the appropriate embedment of mass, energy, space and time. Therefore it means the Vedic science says that the universe is composed of matter (Shiv) and space (Vishnu) which is operated is a balance phenomenon by Adi- Shakti at a cosmic time scale (Brahma). However, this is applicable for the concept of a *Multiverse*. So it is not a question that the universe is made up of Quarks, electron, Proton or the positron because all these are the parts of three fundamental dimension of Mass, Energy, Space and Time.

It is to be realized that Matter, Energy, Space and even Time in its uniform is Par-Brahmor Shiva.

2.4 Relativity

Main article: Theory of relativity

Before **Einstein's** work on relativistic physics, time and space were viewed as independent dimensions. Einstein's discoveries showed that due to relativity of motion our space and time can be mathematically combined into one phenomenon — space-time.

Einstein's Vedic view is that Time (Brahma) and Space (Vishnu) is not the separate dimension but as a single dimension that means Brahm is the only true, rest is illusion.

In sooth, Brahma, Vishnu and Mahesh (Rudra) are one, performs various tasks with the help of Adi-Shakti a singularity.

It turns out that distances in **space** or in **time** separately are not invariant with respect to Lorentz coordinate transformations but distances in Minkowski space-time along space-time intervals are which justifies the name.

In addition the Time and Space dimensions should not be viewed as exactly equivalent in Minkowski space-time. One can freely move in space but not in time. Thus the Time and Space coordinates are treated differently both in **special relativity** (where time is sometimes considered an **imaginary** coordinate) and in **general relativity** (where different signs are assigned to time and space components of space-time metric)

Space is Vishnu; Time is Brahma (Viranchi).

2.5 Cosmology

Relativity theory leads to the cosmological question of what shape the universe is and from where does the space came? It appears that space was created in the Big Bang, 13.8 billion years ago and has been expanding ever since. The overall shape of space is not known, but space is known to be expanding very rapidly due to the Cosmos.

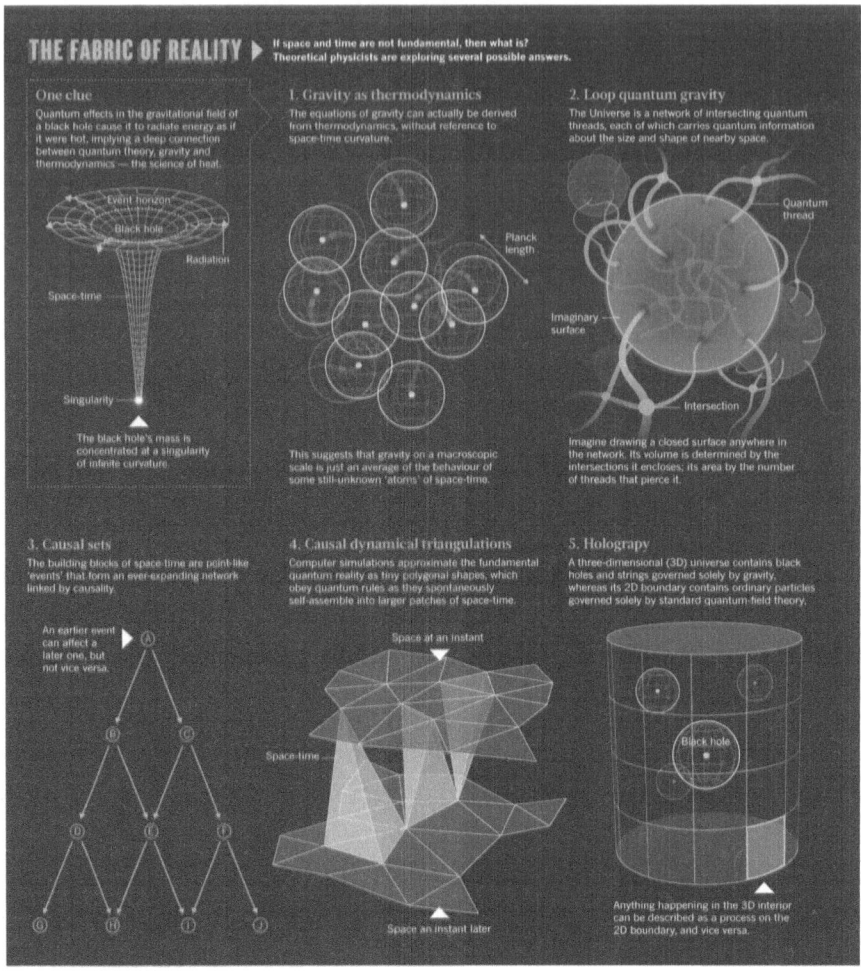

For this, the answer is Space, which appears after the big bang was there even before the big bang but that lord Maha – Vishnu i.e. Big Space was in dormant stage within the Singularity, that means in Sanskrit it is said that he is in **Yog-Nidra**.

Just by the cause of Big Bang, a cosmic ray of light appears from the singularity which influences the appearances of Mass and then Space. And the rapid expansion of space due to cosmic inflation has already described in Shrimad bhagvatam story.

Though there were many distinct stories of shrimad bhagvatam, Shrimad bhagvat Gita, Devi Bhagvatam and other Puranas narrated by

Shri Vedvyasa, but are proven by seers and if we apply our intellect then we can see the findings of modern scientists is same in coordination to Lord Maha Vishnu is precisely the Space in boundless dimensional extent in which, objects and events have relative position and direction.

Shital:- I want to know more evidently, with all the aspect of modern and Vedic science? Ethically, I believe as you said that Lord Vishnu is truly the Space and Brahma Ji is the Lord of Time, but is there any references available in Vedas, Puranas or in Upnishads in the view of these facts or otherwise what you said is again considered a myth for future generations?

Vijay:- I intend to elaborate science to produce a useful model of reality, So now I am going to explain to you complete science, listen attentively; **I shall decode all the myths into realistic science with proper documentation and references of Vedic, Puranic and modern astronomy.**

Its Irony, keeping Sanskrit reserved with limited scholars of India and subcontinents and could not recognize as a universal language.

Those who know Sanskrit in the modern era do not know the science and those who know the science don't have the knowledge of Sanskrit.

In fact, Sanskrit is the only language of ancient scientific theories coded within it, if Sanskrit recognized as a universal language, then Vedas will be the record of perfect documentation in the support of modern science which are the deep ocean of scientific knowlwdge.

Listen carefully I will uplift you step by step with more advanced in fact ancient anecdotal science.

In the holy city Kashi, once upon at the beginning of Kal Yuga the conference was organized by the scientists, scholars and the spiritual masters. This meeting brought together scientists from diverse disciplines working in various parts of the world having the common goal of understanding the universe, cosmology, astrophysics and possible new discoveries and finally in the summit to attain salvation.

In that era, the common goal is to give the universe, the wisdom of supreme science.

The Ancient Sages the seer scientists started to participate and assemble from all over the Jambhudeepa. The aim was to create a platform for attaining spiritualism, to have the opportunity of listening to latest discoveries in the field and sharing results and ideas. It was hoped that the meeting will bridge people to initiate new collaborations between mankind.

The distinguished speaker was Suth Ji, the son of Vedic-Vyasa Ji and the audience was the scholars called Sonakadi Rishi.

This *meeting* brought observers and theorists *around the world* together to discuss the chemical and astrological evolution of the Universe *across the* entire sweep of cosmic history.

Our topics will include the same what was narrated by Lord Brahma ji to Devrishi Narad Ji: Nandishwar Ji to Brahma putra Sanat Kumar Ji, From Sanat Kumar Ji to Vyash and Vyash Ji to Suth Ji and from Suth Ji to Sonakaadi Sages at the beginning of our era, Kal Yuga:

The birth of the first stars, the formation of the first galaxies, the formation and growth of supermassive black holes, galaxies at the peak of star formation in the universe studies of resolved stellar populations and metal-poor stars in the Milky Way and nearby galaxies.

Shital:- Devrishi Narad Ji is known as the ocean of Knowledge, If he wants to know anything about then that might be definitely to enlighten the world.

Vijay:- Devrishi Narad is although preceptor, but remember "You may know many things, but you do not know everything. The primordial knowledge comes from a very long time certainly from the origin of the universe.

This is the story when Devrishi Narad prowl in the search of the Eternal knowledge. There he learned everything about the divine origin of the universe, all about the powers of soul and spirit. Vyash Ji narrated the scared science to his son Sutji,

Vyash Ji Said:

विश्वोद्भवस्थितिलयादिषु हेतुमेकं
गौरीपतिं विन्दिततत्त्वमनन्तकीर्तिम् ।
मायाश्रयं विगतमायमचिन्त्यरूपं
बोधस्वरूपममलं हि शिवं नमामि ॥

जो विश्वकी उत्पत्ति, स्थिति और लय आदिके एकमात्र कारण हैं, गौरी गिरिराजकुमारी उमाके पति हैं, तत्त्वज्ञ हैं, जिनकी कीर्तिका कहीं अन्त नहीं है, जो मायाके आश्रय होकर भी उससे अत्यन्त दूर हैं तथा जिनका स्वरूप अचिन्त्य है, उन विमल बोधस्वरूप भगवान् शिवको मैं प्रणाम करता हूँ ।

वन्दे शिवं तं प्रकृतेरनादि
प्रशान्तमेकं पुरुषोत्तमं हि ।
स्वमायया कृत्स्नमिदं हि सृष्ट्वा
नभोवदन्तर्बहिरेष संस्थितो यः ॥

मैं स्वभावसे ही उन अनादि, शान्तस्वरूप, एकमात्र पुरुषोत्तम शिवकी वन्दना करता हूँ, जो अपनी मायासे इस सम्पूर्ण विश्वकी सृष्टि करके आकाशकी भाँति इसके भीतर और बाहर भी स्थित हैं ।

वन्देऽन्तरस्थं निजगूढरूपं
शिवं स्वतस्त्वमिदं विचष्टे ।
जगन्ति नित्यं परितो भ्रमन्ति
यत्सन्निधौ चुम्बकलोहवत्तम् ॥

जैसे लोहा चुम्बकसे आकृष्ट होकर उसके पास ही लटका रहता है, उसी प्रकार

Shiv the Matter (Mass of the Universe)

Vijay: Narrating on Shiva is similar to being speechless but more specifically in regard, I got the ability because of his divine grace, I may able to say something about the scientific natural phenomenon of Lord Shiva.

Before embellishing Lord Shiv, I salute to Lord Shiva by his grace I could able to describe him. Otherwise, Shiv cannot be even imagined by gods. The Shiva is the only reason for origin, existence and continuation of the entire world. He who is Maha-Kaal the everlasting, the one who is the consort of Umma Gori the divine energy Adi- Shakti and (Vidit tatbh anantam kirtim) the known endless **Matter** without beginning.

Shiva who along with **Maya (a form of energy as Adi Shakti)** but never indulge to her illusion, I salute to Lord Shiva and Parvati (Adi Shakti) in combine form.

This primordial Mass in the aspect of Shiva is also called to be Supreme Being (Parmatama), God, Sambhu, and Maheshwara or Brahm or Brahman. Shiv holds entire of the celestial bodies, galaxies and all the matter of the universe.

There are two words Shiv and Shav, Shiv is Ishan (Ishwara the Lord of Life) and Shav (Non-Living Particle) also called corpse and carcass i.e. death body. So the particle or matter assembles together with Soul or Space (Narayana) forms a living entity.

Thus entire universe existence is adynamia (without life) without Shiv, Shakti, Brahma and Vishnu. Same proven by modern science which says this world could not be there without Mass, Energy, Space and Time.

The divine Lord Shiva created whole the world and is placed everywhere in the Space (Maha Vishnu) within and out of himself.

Thus God Vishnu and Shiva are very equal and should not be comparable because Mass, Energy and Space are directly proportional to each other is one Brahm.

It is never the coincidence that Shiv and Shakti are the coincide, known as mass and energy balance.

Chapter 3

Story of Big Bang and Big Crunch, Origin and Dissolution of Universes

Shital:- It is the era of Kal-Yug, so many principles and theory put forth and many are yet to come, to understand the cosmos, but sometimes those theories are confusing too. There is no one to prove what was there before the Big- Bang. Is there any Vedic Scientific theory, which is evidently excathedra, the state before the Big-Bang?

Vijay:- Once Seer Narad asked Brahma with the same curiosity like you.

Narad asked:- Father, Now It is clear to me how the universe comes in existence, but I am very curious to know that what was there before the Maha-Naad or Anhad Naad (Big-Bang)?

Lord Brahma Says:- Hey Son, your are always be very active in the welfare of mankind, I know that you are raising this for the sake of people and generations one after another will practice this for scientific researches. You use this knowledge by delivering it to the holy hearted human, always remember that any research should *always* aim and centralize at the *welfare of humankind*. Listen, the secret of eminent Shiva (Aghan – 'Non-Danse' state of mass) Shiv-Tatva (Mass Particle) which I shall narrate to you now.

The primordial sound (Naad) can be realized and experienced even today when the yoga enters into deeper states of trance, the awakening Kundalini begins to course through the body, the subtle body is activated and the brain experiences a reverberating natural silence of Naad and Bindu.

Hey Narad, The Yogi's experiences a sense of purity, rejuvenation and alertness within. At this point, one may hear subtle sounds in the ear. The sounds which the Yogi's hears tend to vary depending on the inner plane of consciousness to which one is currently attuned.

सारा चराचर जगत् बिन्दु-नादस्वरूप है। बिन्दु शक्ति है और नाद शिव। इस तरह यह जगत् शिव-शक्तिस्वरूप ही है। नाद बिन्दुका और बिन्दु इस जगत्का आधार है, ये बिन्दु और नाद (शक्ति और शिव) सम्पूर्ण जगत्के आधाररूपसे स्थित हैं। बिन्दु और नादसे युक्त सब कुछ शिवस्वरूप है; क्योंकि वही सबका आधार है। आधारमें ही आधेयका समावेश अथवा लय होता है। यही सकलीकरण है। इस सकलीकरणकी स्थितिसे ही सृष्टिकालमें जगत्का प्रादुर्भाव होता है, इसमें संशय नहीं है। शिवलिङ्ग बिन्दु नादस्वरूप है। अतः उसे जगत्का कारण बताया जाता है। बिन्दु देव है और नाद शिव, इन दोनोंका संयुक्तरूप ही शिवलिङ्ग कहलाता है। अतः जन्मके संकटसे छुटकारा पानेके लिये शिवलिङ्गकी पूजा करनी चाहिये। बिन्दुरूपा देवी उमा माता हैं और नादस्वरूप भगवान् शिव पिता। इन माता-पिताके पूजित होनेसे परमानन्दकी ही प्राप्ति होती है। अतः परमानन्दका लाभ लेनेके लिये शिवलिङ्गका विशेषरूपसे पूजन करे। देवी उमा जगत्की माता हैं और भगवान् शिव जगत्के पिता। जो

The evidence from Shiva Purana

Vijay:- The seers (Ancient Scientists) in their experience given us exactly the clear wisdom of cosmic truth; they found psychological evidence in their observation in which there are two cosmic realities, the persistence and the change (evolution).

The two primary fundamentals revels being and becoming, the Vedanta says it as Mula Prakriti and Prakriti. Mula Prakriti is the Potential energy is the state of being, Mula means the root of creation and from that same or by its power everything emerges. So that is the process of becoming a Kinetic (energy) form of nature from the static form. The world which is moving around us is Prakriti.

Before Maha-Naad there was Nishkal Shiva, the Shiva is the name, conjointly used for Adi-Shakti and Lord Shiv. We found in Shiv Puran which is narrated by Nandishwara to Sanat Kumar Ji, same was narrated by Sutji to other Sonakadi Seers.

Sutji Said:- Hey Sonakadi Rishis, before Maha Naad, Nishkal-Shiva desired with pure consciousness (Chit) for the creation, however the cosmos was created by the action of Adi Para Shakti or it's by an energy called Maya. Indeed Maya is energy, which is having a capacity of transformation. So in short, we can say, Nishkal Shiva transformed to Sakal Shiva, Same as Devi Shiva or Mula Prakriti or Shakti transformed into Maya Devi.

Sakal Shiva or Nishkal Shiva is one with dual phenomena. It's the Shakti in its Mula Prakrit.

A non dense state (Aghan) of Nishkal Shiva was then turned to an Individual form of energy of high density called Sakal Shiva.

This is Bindu Naad Swaroopa and the form is called Omkara, which is indeed a primordial Sound.

Vijay:- Brahma Ji said to Narad, The whole universe (Charachar Jagat) was the form of Singularity, a String **(BINDU NAAD-SWAROOPA).** Singularity is energy (Bindu Shakti), Naad is a string. The entire cosmos emerged from this Singularity. This world is the image of Bindu-Naad i.e. Shiva-Shakti.

There were three states proceeded one after another **Aghanavastha** (the Non-Dense state), **Ucchanavastha** (The highly noble condition), **Upayogavastha** (ophelimity or the state of usefulness) was one after another stages want through the **Kali,** the Dark Energy because before the Maha-Naad there was only darkness. That darkness is the Goddess **Kali.**

These are the three primary states of transformation of Nishkal Shiva, the energy from highly potential state in to different states.

The Shiva-Lingam represents the same that how this universe emerged from Naad Bindu.

Naad and Bindu are one and the same the Singularity (Shakti), being the names of two of her states which are considered to represent as being more prone to creation (Becoming) **Ucchanavastha.** The states of Singularity i.e. Shakti-Bindu suitable for emerging of creation of this state is called in Sanskrit (Upayogavastha- the suitable condition).

Vijay:- The same secret was narrated by Nandishwar Ji to Brahmaputra Sanat Kumar Ji.

Shital:- Sanat Kumar is one of the genius, divine, spiritual Manas Putra of Lord Brahma. He is one among 1st four sons of Viranchi Brahmaji; he is the spiritual master why he seeks that knowledge from Nandishwar Ji. Can you explore that hidden fact of Puranas?

The Beginning of Epopee

Vijay:-There was the ancient group of scientists, In Vedic language; they are called Sages (Magarishi's or Rishi's). In the modern world, there is a realistic error in the understanding of people's. We have to comprehend the contribution and discoveries given by those sages for the world by their research.

From the description of Shiv Puran, I will narrate you the same science which was explained in the primeval period by Lord Brahma Ji to Narad Muni and after the cyclic epoch this scientific knowledge was delivered by Nandishwar to Sanat Kumar Ji.

But there is an interesting anagogic anecdotal story; it's the science of Kashi, the scientific phenomena of Lord Vishwanath and the essence of cosmological truth.

Today we see the temple stands on the western bank of the holy river Ganga and is one among twelve Jyotirlingas, the holiest of Shiva temples. The main deity is known by the name Vishvanatha or Vishveshvara, which means *the lord of The Universe*. Varanasi city is also called *Kashi*, and hence the temple is popularly called Kashi Vishvanath Temple.

On those days the Viswanath Lord Jyothi Lingum was in open ground. Temple was not build on those days, in reality all the primordial Jyothirlingas are spontaneously emitted and are called Swyambhu (स्वयम्भू) "self-manifested" or "that is created by its own accord". These temples at the initial period are wall-less and roofless.

Nandishwar Said:- Hey Sanat Kumar Ji, Once the group of sages took holy bath in river Saraswati and performed their worship and rituals. After that, they proceeded towards Kashi. At Kashi, after taking bath in river Ganga they visited, viewed (darshan) and had worship of Lord Vishwanath.

The sages gathered at Kashi in the summit of a scientific conference. They saw extremely radiant effulgence appearing in the sky, in which they saw thousands of sages who had accomplished Pashupat Vrata getting merged. The radiant effulgence vanished in no time.

The sages were very curious to know about that radiant effulgence, so they went to Lord Brahma and asked him about it. Lord Brahma told them that the radiant effulgence appeared in the sky, was to inspire the sages so that they may accomplish the Pashupat Vrata to attain Salvation. Lord Brahma then instructed them to go to the Sumeru Mountain where I (Nandi) was supposed to teach the methods of doing Pashupat Vrata to Sanatkumar.

Shital:- Even I am too curious, to know about that radiant effulgence? Was that some miracle, illusion or is it pragmatic science?

Vijay:- Whatever we saw and see that everything is the part of universal science of the Almighty Shiva, whole worldly phenomena comes in front of us in a routine cyclic order but if something comes suddenly we wonder it as a miracle because we are not use to that event.

That radiant effulgence which sages the scientists of that era saw was the **cosmic microwave background (CMB)**, which was electromagnetic radiation.

Shital:- Oh that's amazing, you mean the CMB, the CMB which is the oldest light we can see and having its presence long before the solar system or even before the birth of our galaxy.Wow, interesting can you please explore more; I am contemplating to know about this?

Vijay:- The sages saw the radiant effulgence was the **Cosmic microwave background (CMB)** is electromagnetic radiation left over from an early stage of the universe after Big Bang (Anhad Naad). It is the residual heat of creation from Anhad Naad. CMB is a faint glow of light that fills the universe and the representation of Lord Shiva's Jyothirmay Lingum from which the universe evolved, this light set out on its journey more than 14 billion years ago, long before in the very early phases of the universe.

In modern science the scientists **Arno Penzias and Robert Wilson** detected the **cosmic microwave background (CMB)** in 1960 for the first time but Indian Ancient Seers (Scientists) knew from the history of the Vedic scientific era.

This discovery is important in the support of Big bang, same way the radiant effulgence is the leftover of Anhad Naad.

Shital you should also note that Brahama Ji has, mentioned that those who perform Pashupat Vrata they will attain salvation or able to merge into radiant effulgence.

Shital:- Can you clarify what is the relation between salvation and merging into radiant effulgence which you described very well as a **Cosmic microwave background (CMB).**

Vijay:- First you understand Pashupat, Pashu means something like thunder or radiation, Pashu also means by which something tides or the binding phenomena.

In the material world, our soul is bounded by our material body. Everything in this world is the bundle of different types of bounded materials. Likewise, our cells are bounded by tissues and group of tissues make our body. Same way our embodied soul which on evolvement always remains bounded by our material body, the entire world is the embodiment by unified forces, dependent on material objects.

In Pashupat, Pat means shoot or strike and when a shot object hits the target, the targeted objected get destructed. So when Pashupat was performed on this material body, our soul gets liberated from material body.

Lord Shiva is the Pashupathi the god of liberation when the sage's perform Pashupat Vrata they attained salvation and the liberated souls of sages got merged into the radiant effulgence of **cosmic microwave background (CMB).**

Lord Shiva uses his **Pashupath** for creation and destruction, good and evil, to bound and liberate. When he shoots Trident on the evil or the devils they get destruct and when Pashupat performed on Sages they attain salvation.

In the beginning of the universe how the cosmos got radiant effulgence the **Cosmic microwave background (CMB)** you will come to know in our further discussion as we proceed in this scientific tale, you will get shower with this secret knowledge and lamina of your brain will be decoded.

Now let's come in the story, when Nandi came to teach Pashupat Vrath to Sages on Sumeru Mountain. But before this anecdote there was an incident, listen.

Once upon in early epoch, **Sanatkumar** the Manas Putra of Brahmaji became very arrogant from his solitarily. One day lord Shiva arrived at his place but Sanatkumar did not get up to greet him. At this Nandi became furious and cursed him to become a camel. Sanat Kumar got transformed into a Camel.

This was done by the processes of complete mass transformation; the mass of Sanat Kumar Ji was transformed by Nandishwar Ji into camel through molecular cloning and transferring genetic material. In modern era it can be done by either heat shock or electroporation. Same way it was done by Nandishwar ji through his Tapas (heat shock).

Lord Brahma worshipped Shiva to liberate his Son- Sanat Kumar from the curse of Nandishwar. Lord Shiva became pleased and blessed Sanat Kumar, as the result of which he regained his Divine Deity body and rejoined as a spiritual master along with his three brothers.

After regaining his deity body Sanat Kumar commenced a tremendous penance. Lord Shiva instructed Nandi to go to Sumeru Mountain and preach Sanat Kumar.

Sumeru Parbat is a mountain in the Gangotri Glacier region of Himalaya in Garhwal, Uttarakhand, India. It is 6350 meters high. It is encircled by Kedarnath & Kedardome in the north, Kharchakund in the west & Mandani and Yanbuk in the south.

The sages reached Sumeru Mountain as per the instruction of Lord Brahma. They saw many sages meditating at the bank of a pond. They also saw Sanat Kumar engrossed in his meditation at a little distance from the other meditating sages.

The sages went near Sanat Kumar and told him about their penances done for ten thousand years. They also revealed to him about the purpose of their arrival. Right then, Nandi arrived there, accompanied by his ganas Sana Kumar and all the sages welcomed him. Sanat Kumar introduced all the sages to Nandi and told him about their tremendous penance done for ten thousand years. Nandi became very pleased and blessed them. He then preached Sanat Kumar and all the sages on the Shiva tatava.

Vijay:- Nandishwara preaching of Shiva tatava was started and the same we called String theory in modern physics which is as:

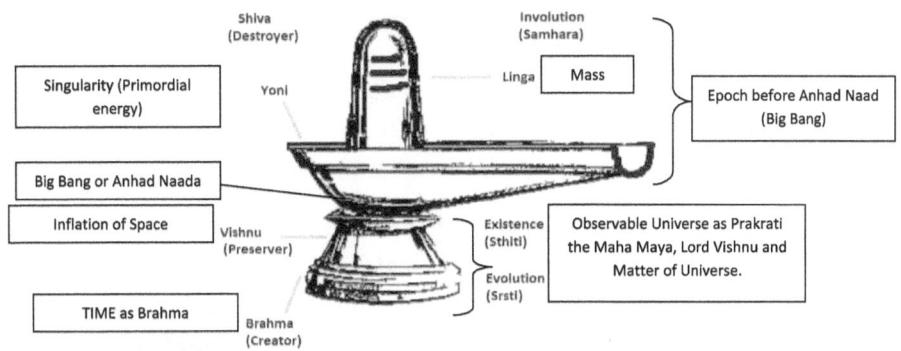

Evidence of cosmic events that happened before and after the Big Bang (Anhad Naad) in the view of Shivlingam

3.1 Primordial Epoch (Adi Kaal) - String Theory of Origin of the Universe or Vedic Theory of Naad and Bindu

Nandishwar Ji said:- Hey Brahma Putra Sanat Kumar Ji, Nishkal Shiva is beyond any observation, therefore Lord Shiva is beyond the worldly identification of gender, caste and race.

There is no mass (Ghana) in Nishkala Shiva, that is the state of Brahman before Maha-Naad, represents the Aghanavastha (the massless state). So the gross world is the morph of Shiva-Shakti swaroopa.

Naad and Bindu are one and the same, Shakti (Singularity), Shiva-Shakti is one state of dual phenomena, Naad is sound and Bindu is the point of energy from where the sound produced. So it is the sound energy which influences the creation.

This Naad (sound) arise from the vibration of energy. Vibration or string is the effect of potential energy and caused Naad. This state was promotional energy suitable for a creation (Ucchunavastha, the eminent state). Under the influence of Naad, Bindu achieved a suitable condition (Upayogavastha) to produce a creation.

Up to the stage, there was no mass (Ghana) so this is called Nishkala Shiva, Nishkala Shiva is the united form of Naad-energy is a Brahm Aghanavastha (Non-Dense state).

Vijay:- Is there any mismatch or disproportion in this science of ancient Vedas with modern scientific theory, I think no, actually it's more advance then that of modern science, modern scientists can't able to prove anything before big bang because Nishkala Shiva or Singularity is not a subject or object to come in any experiment and observation.

Nishkal Shiva or Adi-Shakti can be realized only with devotion and beliefs. The contribution of Ancient Seer's are outstanding, they have sacrificed their lives in penance and mediation to understand the reality of Parbrahm. We should never forget the ancient Sages, who dedicated their lives to bring happiness for today's generation.

The Nishkal Shiva - Shakti (Energy) transform themselves to Sakal Swaroopa which is a Sat (Truth) and from Sakal Swaroopa emerged Atman (Chit) and the feeling of Anandam can come only from SAT CHIT (True Atman or the pure Soul) i.e. Sachidananda (Sat+Chit+Ananda). Chit is the pure consciousness and Ananda come only with true consciousness. The first Soul Purusha, by the wish of Shiva-Shakti achieved inflation and become the Space (Maha Vishnu).

For Brahm, we cannot use the identification of gender like Male or Female, he or she; the supreme Brahm is a united form of Shakti and Shiv. That's by Shiva first appeared in front of Shri-Vishnu and Brahmaji in **Ardhnarishwara** swaroopa (Phenotypic character or androgynous). Ardhnarishwara means Ardh (Ardh + Nari + Ishwara) the **"supreme being who half Isa (Nar) and half women (Nari)"**.

There was darkness all around after the last Big-Crunch and before the Maha-Naad or Anhad- Naad (Big-Bang) there was nothing (The theory of Nothingness).

This darkness can be called as Maha-**Kali (Dark-Energy)**, Kali is one form of Maha-Vidya (Energy Form) out of ten Maha-Vidya, Maha means Great and Vidhya means wisdom.

Again the description of **Nasadiya Sukta of Rig-Veda** in **Rudra-Samhita** of Shiv Maha Puran comes here.

Vijay:- Brahma Ji said, Hey Narad, After knowing the wisdom from Nandishwar, Sanatkumar passed this Pashupat knowledge to

Sage Vyasa, who again passed it to Sutji. Thereafter, Sutji revealed this knowledge to those sages who had assembled at Prayag (Kasi). After receiving that knowledge from Sutji, all the sages went to Prayaga tirtha and took the holy bath.

As they saw the signs of approaching Kali Yuga, they went to Kashi and performed the Pashupat Vrata. All of them attained liberation with the blessings of Lord Vishnu.

Listening to Shiva Purana for one time liberates a man from all of his sins. Listening twice helps human being to attain devotion in lord Shiva. Listening to Shiva Purana for thrice attains to the abode of Shiva.

Shital:- The Sages attained liberation means they merged with the Cosmic Microwave background radiation, isn't it?

Vijay:- Yes off course, remember the salvation is merging of Jivatma with the Parmathama and this is the state of attaining riddance from sorrow and grief and finally to get purge from the cycle of birth and death.

So the Cosmic Microwave Background (CMB) is the primordial light of 13.5 billion years back. It means the soul of sage's merged reversely with the Shiva and attain SAT CHIT ANANDAM.

Shital:- In modern science **Cosmic Microwave background** may be a simple the radiation leftover of Big bang. It's really great to know the wisdom of Pashupat, which reveals the secret science of Cosmic Microwave background (CMB) and CMB is the radiant leftover of Maha Naad or Anhad Naad for which it's not possible for an ordinary being to recognize the reality of this cosmic truth.

Vijay:- Exactly, You are right, the radiation of **Jyothirmay Lingum** appeared immediately after Maha Naad or Anhad Naad is Cosmic Microwave background (CMB). Brahma Ji further said: To understand Pashupat, you should know the origin of Pashupath, Pashupat is therefore from the beginning of the cosmos.

Brahma Ji said:- Hey Devrishi Narad, The Glory of **Shiva Tatva** is a paramount, very auspicious, topmost and wonderful in itself. When the whole (Charachar) world was lost, everywhere darkness, the only darkness was wrapped. The day and nights don't have any meaning because neither there was Sun nor Moon, nor there was Stars, Planets, or the Galaxies and there were no constellations. Earth, Water, Air and Fire were also not found. Not was the Sound and Senses haven't perceived, also there were no shape and nor an *odour*.

Macro, micro, major smaller doesn't have any mean, there was only almighty Nishkal Shiv-Shakti without the Soul Space (Remember lonely sky means It's emptiness). This way there was only darkness everywhere.

Hey **Narad** what was only found to be there was only **TAT SAD "Brahm"**, these were the characters of ethereal Brahm. In Tat Sad Brahm, Tat is particles or concentrated Mass, Sat is the Spaceless Soul and Brahm is the Divine Energy coupling the Mass and Soul in the form of Aimkari **(Singularity)**.

This and that, here and there also doesn't have any mean. The only mean is the SAT the real truth, which the Seers (Ancient Scientists) found in their meditation, though it cannot be described. That Truth, Knowledgeable, Infinite, Baseless, Shapeless but the Single reason for all is the TAT SAT or the Brahm.

That SAT is not the subject of thoughts, It never grows and nor ends, it is neither scanty nor lengthy. Thus it cannot be described in words.

Hearsay (Sruthi) (The Secret code) the Revelation, sacredly only says that नेतिनेति (न इतिन इति) means "not this, not that", or "neither this, nor that". Hearsay (Sruthi) (The Secret code) the Revelation, sacredly only says that नेतिनेति (न इतिन इति) means "not this, not that", or "neither this nor that".

Because In Vedic scriptures it is said that "Parabrahmam" or "The Ultimate Truth" cannot be seen, felt, experienced or understood by anyone by any means.

तत्त्वमस्यादिवाक्येन स्वात्माहिप्रतिपादितः।
नेतिनेतिश्रुतिर्ब्रूयाद अनृतं पांचभौतिकम् ॥

Then the Supreme Brahm desired to create infinite and multiple from the Singularity, that finite desire to be infinite, from dimensionless to multi-dimension irrespectively.

Brahma:- That Brahm is truly the combined form of all the material, non-material, ego and spirit, a genus of all the species of today's universe. All the souls I, you and all that we look today that all was there within the Brahm. Brahm (ब्रह्म or शिव) is the originator of Time as Brahma (ब्रह्मा), Space as Vishnu (विष्णु) and Matter as a Mahesh or Rudra (रुद्र शिरोमणि) or Shankar. He is also addressed as Mahabrahma (मंहब्रह्मा), Mahavishnu (मंहविष्णु), and Mahashiva (मंहशिव) in another word Singularity (Adi-ParaShakti).

Narad:- Then what happened, father?

Brahma:- Divine Brahm is sedate **Sada-Shiv** which is having all the qualities completely characterized universally versatile and available.

In the process Bindu and Shiv (Bindu the Shaktim and Shiv the Tatvam) is the substratum of all, that there was the only Singularity enveloping whole Mass created the universe.

Om Tat Sad was the ultimate truth by the moment of the big bang, with sound OM the first particle TAT has been appeared as (Sada-Shiv) Mass along with its Adi-Shakti (Energy).

As all you know Big bang occurs from a **singularity.** Singularity is the Adi-Parashakti posses Mass inside her. **Adi-Parashakti** is her formless state which is beyond to this temporary cosmos and **Adi-Shakti** is her omnipresence and ostensibly forms of all energies.

Narad:- Oh Lord, Your oration is an attribute and virtue of fortune which cannot be described.I want to know more about how that Singularity appeared as a empyrean goddess, what was her shape and how she looks like?

Brahma:- From Sada-Shiv or Adi-Parashakti a Angelic beautiful Gorgeous self-oriented Shakti **Bhuvaneshwari** emerged and she wears entire of the universe within her.

3.2 Indivisible Shiv and Shakti and Theory of Mass and Energy Balance

Shital:- Yes Vijay, **Big Bang** is a model of creation, that explains how the universe begins, but it's also true that is still a puzzle for modern scientists, is very scientifically explained in Vedas. Vedas says that Shiv is the only cause to bring the universe forth. Can you explain it in a more scientific way supported by your Brahm Science?

Vijay:- Modern Scientific support:-However, in the beginning of the universe, the first 'Kalp' or 1st epoch (Kalp means eon), with in the divine dark energy (Devi Kali) which exist as a singularity (Brahm the Nishkal Shiva). *Anhad Naad* **begins with the burst having a huge amount of mass and energy which are Shiv and Shakti.**

Now, a team of physicists says the Big Bang should be a model as a phase change: the moment when an amorphous, formless universe analogous to liquid water cooled and suddenly crystallized to form four-dimensional space-time.

Mulaprakriti is the form of Singularity as the material and instrumental cause of things is that potentiality of natural power which manifests as the Universe.

To understand it in a current scenario, we take the help of modern scientific vocabulary. There is a misconception that the Sanskrit language is only a language for chanting mantras in temples or religious ceremonies. Sanskrit words have the same meaning as modern astronomy. Words that are mostly fitted in today's world of scientific society. That is the reason I have used certain world like Big-Bang as (Maha-Naad or Anhad Naad), Singularity as (Aimkari) a Space as a Vishnu Ji, and so on.

The first is the `Spirit or Purusha or Brahman and Atman (ParmAtma) which is a limitless Spirit (Sat), Consciousness (Chit) and Bliss (Ananda).

So who is this Sat-Chit-Ananda **(Satchitananda)**?

He is Lord Maha Vishnu called as Sachidananda (मंहाविष्णु) because he is **Sat** means the real Truth, **Chit** means of pure consciousness and when there is only truth (SAT) in the consciousness (Chit) that state is called **Ananda** (The extreme happiness). We can say the state of happiness or Bliss from all Side. This is the Phase of inflation of space and there is only Pratham Purusha Lord Vishnu blessed by **Anandam**.

Vijay:- Now further we will find the Vedic description of Mass and Energy, here the question is. Can Energy exist without Mass?

We will find out the Vedic description briefly explained by Prajapati-Brahma to Narad Muni?

Arupah kevalah svastho mahadevah svabhavath [Kurma Puran – 1.10.82]

This clearly says that Mahadeva or Shiv by nature is formless (Arupah) and Single.

Amorphous or Formless (Nirakar) or Atma, the Universe means truly TAD SAD Brahm. The Mass in Singularities leads to Phase transition by a burst of huge energy, therefore its Mass which brought very first forth the Space and then Time. This Mass is the compressed form of cncrgy. That means Shiva and Shakti are not the two entities in spite they are one as Shiv-Shakti.

The Energy is never be separated by Mass hence Shiv and Shakti is the two side of one coin. During this phase, Mass and Energy were inseparable.

Here also establishing the relation between Sada-Shiv and Adi-Shakti which can be scientifically called the *Theory of Mass and Energy Balance*.

According to Aitareya Upanishad, from Atma (Nirakar Shiv) evolved Amba (Energy). Amba exists beyond the heavens and is glowing in the form of Adi- Shakti. From the same essence as that of Amba, the Purusha (Space in the form of Vishnu) was drawn out and laid to protect the Universe.

One of the fundamental laws of physics states that mass (SHIV) can neither be produced nor destroyed, but mass can be conserved and same has been claimed by Vedas, Shiva is Anadi and Ajanma (Endless and Unborn) Equally fundamental is the law of conservation of energy (Shakti). Although energy can transform, it cannot be created or destroyed.

In the further story, it will be clearer to you, how the energy change its forms for the weal of the entire cosmos, which is now scientifically claimed in Kali Yug.

These two laws of physics provide the basis for two tools which are used routinely in environmental engineering and science, the mass and the energy balance.

Now listen, what Brahma Says to Seer Narad which is going to break the most hidden secrets of the universe for today's generation and its records are decoded in Vedas.

Brahma Said:- Hey Narad

Then there was an anecdote, with a short of Vibration in the Bindu (Singularity) with a mass (Shiv as Tatvam) result in tiny strings that formed loops were the beginning of the universe.

The **Adi Shakti** is called Divine Soul Energy, Nature, Knowledge and Wealth also called the mother of trinity that is Brahma, Vishnu and Mahesh.

She is also named as Ambika, her sheen shines like thousands of moon shines together. She is omnipresent, She is the only reason for all causes and sources; her eight arms represent eight dimensions. She is the seraphic feminine itself, she was the something called nothing - 0 – (Soonya) becomes everything from that single when it wiggles and twists to form - 8 - Infinite life. Soonya or Zero means that which is out of our perceptions out of the imagination of anyone. It doesn't mean it was absolutely nothing in spites it holds everything of whatever can be believed in universe. We ordinary humans can't know the actual reality.

In fact in reality her presence is in the form of Adi-Parashakti as a singularity.

The Goddess Adi Para Shakti in her deity form holds many weapons with divine ornaments. She is one in all and all in one. She is also called the mother of trinity i.e. Matter, Space and myself as a Time.

The Adi Parashakti is Mulaprakriti which always exists, which means "the source of life", from which everything emerges, which flows from the Mahadevi (aka Adi Parashakti or the Singularity), a divine feminine matrix. The Goddess Adi parashakti is therefore the primordial energy that governs the universe.

Remember **Narad**, Shiv (Mass) is also called as **Shambhu** (Self originated by own will) that Rudra who is having five faces with three eyes on each and ten arms wearing Trident as weapon and ash-smeared is the one aspect known as the Mahesh or Shankara.

Shiva-Rudra depicted as creater of matter, antimatter and fundamental
forces of the universe

Chapter 4

Inflation Epoch - The Inflation of Space or the Expansion of Maha Vishnu

Shiv and Shiva in Vedas, Shiv means Devadi Dev Mahadev and Shiva is Adi-Shakti (Mass and Dark-Energy)

Lord Shiv-Shakti thought to produce something like inflation to make the universe big and who therefore can take the charge of Universe operation and lead the cosmos.

Hence Maha Vishnu was begotten and expands throughout as the space in all directions.

Then that Mass (Shiv) and Energy (Shiva) thought to ostensive a soul by which universe got enlighted is called **Narayan** also named Vishnu because of his Inflation nature. It's Lord Vishnu "Inflation" process causes a vast expansion of space.

He is the eponym, first men (Pratham Purush) of the universe who escalate the cosmos. He is the Ether, he is Space.

Along with the beginning of Inflation (Birth of Vishnu) the Adi-Shakti also transforms herself in one of its form called Yog-Maya in the form of Energy to help the Vishnu.

Thus this energy (Adi Shakti or Prakriti Devi) comes in existence just after the beginning of Vishnu from the common source so she always play the character of Narayana sister in his various incarnation for the liberation of mankind.

Shital:- So what caused this inflation of Maha-Vishnu? Modern astronomers have many speculations as to its nature, but the convincing answer is yet to be found?

Vijay:- It's my attempt to determine the Vedic-Modern model based on astronomical events sagaciously corresponds, which was coded in Vedas to the reality that drove a great deal of research over the decades. At the beginning of the universe, rapidly expanding space was pretty much empty of matter, but it moves lien amount of dark energy, with that the theory of the universe moves on.

Dark energy is the mysterious force and scientists think that it's Dark energy which is driving the universe's current accelerating expansion.

During inflation, dark energy (Yog Maya) made the universe smooth out and accelerates. But it didn't stick around for long.

In the various phase of the cosmic manifestation, in the appearance of world, its sustainability, its growth, its interactions, its deterioration and in its disappearance, the energies having its basic relation with the existence of the personality of Godhead.

Prajapati Brahma Said:- Hey Devrishi (designation of Narad) this is the story of Maha-Naad (big bang), by the origin of Super Soul Vishnu, the Ether by which invisibility turns to visualize the universe. He is SAT from **OM TAT SAT which together called "Brahm"**, SAT means truly Space and the Spirit.

Narad that was the Vishnu of Purush Suktam mention in Rig-Veda states that:

ॐ सहस्रशीर्षा पुरुषः सहस्राक्षः सहस्रपात् ।
स भूमिं विश्वतो वृत्वाऽत्यतिष्ठद्दशाङ्गुलम् ॥

sahasra sirsa purusah | sahasra aksah sahasra pat |
sa bhumim visvato vrtva | atyatisthad dasangulam || 1 ||

By his appearances only, all the laws of physics got accelerated. He is the one who accomplishes the world. He is an Aglow and Aeruginous (Neelhari) because of his universality and expansion, Lord Shiva named him **Vishnu** (means Vyapak). He is truly Space the ample almightily God.

That Supreme Being is having thousands of heads, the thousands of eyes and thousands of feet, He has appeared as Space (physically a boundless Space).

Pervading the cosmos in all directions, He himself is the Space. Thus he is the first being (Soul) came into existence.

Vijay:- This is the reference of Rudra Samhita and the secret which is to be comprehended here with the 1st verse of Pursha Suktam is that Lord Vishnu is the effect, while Lord Shiva and Shakti (the gross mass and energy) is the cause.

Lord Vishnu in physical cosmology is the **cosmic inflation**, **cosmological inflation**, or just **inflation**, is a theory of exponential expansion of space in the early universe.

The inflationary epoch lasted from 10^{-36} seconds after the conjectured Big Bang singularity to sometime between 10^{-33} and 10^{-32} seconds after the singularity (Aimkari). Following the inflationary period, the Universe continues to expand, but at a less rapid rate.

Sahasra sirsa Purusha means the lord Vishnu presence is everywhere, in every direction of the universe likely having thousands of heads.

पुरुष एवेदं सर्वं यद्भूतं यच्च भव्यम् ।
उतामृतत्त्वस्येशानो   यदन्नेनातिरोहति ॥
एतावानस्य महिमाऽतो ज्यायाँ श्च पूरुषः ।
पादोऽस्य विश्वा भूतानि त्रिपादस्याऽमृतं दिवि ॥

Purush evedagum sarvam | yad bhutam yac ca bhavyam |
utamrtatva syesanah | yad annena tirohati || 2 ||
etavan asya mahima | ato jyayagus ca purush |
padoasya visva bhutani | tripad asy amrtam divi || 3 ||

That Supreme Being is indeed everywhere, He is the Creature's Space (Soul) and that Space is in every being's soul, He confers all immortality and alive forever before the beginning of Time.

The Vedic Purusha Suktam explains the origin of the large-scale structure (evedagum̐ sarvam̐) of the cosmos in the physical cosmological manner.

Quantum fluctuations in the microscopic inflationary region, magnified to cosmic size, become the seeds for the growth of structure in the Universe. He transcends all in his form as food (the universe). Such is his Glory, but greater still is the Purusha (See galaxy formation and evolution and structure formation). Many physicists also believe that inflation explains why the Universe appears to be the same in all directions (isotropic), why the cosmic microwave background radiation is distributed evenly, why the Universe is flat, and why no magnetic monopoles have been observed.

His grandeur is omnipresent, He is the Supreme Being, All creatures are just the micron part of him and the Purusha is none other than Lord Vishnu who expands ad infinitum.

Brahma says:- That's the universe just after Anhad-Naad (Big-Bang) without Matter and Time. There was big inflation that it was pervaded in all directions in microsecond which has expanded and crossed out a doomed shower of sparks, its limits is out of human perception.

Some scholars believes that, the inflation was found to be from the left side of Shiva that means the left portion of Shiv is UMA Devi, the feminine part of Ardhnarishwara swaroopa.

This Ancient science was supported by the theory of inflation which occurred because of strong energy (UMA) and SHIVA (Mass) induces the cause of Inflation and then known as Vishnu pervaded passim. And hence he is known to us as a Param Purusha.

Hence he has originated from the united form of Shiv-Shakti (Mass and energy) or from singularities (Adi-Shakti).

This Vishnu (Inflation) is the Naad (Bang of Big- Bang). It was needed the universe to come in existence because of the Inflation of Vishnu.

Thus the Par-Brahm (Shiv along with Uma) has created a Universe which has been followed by Maha-Naad (Big-Bang).

The Shiv (Mass) along with Dark-Energy is the Big of Big-Bang. This energy is also called Singularity which causes and holds all the universal matter within her.

The Mass and energy together called SOMA (SO means Shiv and MA means UMA) and that is divine **Brahm (Here SOMA is distinguished by Som the Chandrma).**

It is (Shiv) who takes the form of Vishnu as clearly stated in the below verses of Vedas.

Vijay:- Rig Veda also states the same that Soma (Sa + Uma = Lord with Uma = Shiva) beget (given birth) Vishnu. So that is Mass and Energy together generated the Space i.e. generated Vishnu by the processes called inflation.

> "somaḥ pavate janitā matīnāṃ janitā divo janitā pṛthivyāḥ
> janitāghnerjanitā sūryasya janitendrasya janitota viṣṇoḥ "
> (Rig Veda.IX.96.5)

Rig Veda clearly says, SOMA (Shiv with Uma i.e. Mass along with energy) beget (Janita i.e. given birth) to earth, Celestials bodies, fire, Sun and also Indra and Brahma are the form of Vishnu. Shiv and Uma are the parents of all of them.

SOMA, in turn becomes Aditi and the Aditi gave birth to the Celestial bodies the Agni (Fire) The Adithya (Sun).

> "namo girishaya cha shipivishhtaya cha |" (Yajurveda iv.5.5.f)

> > "Salutations to the Lord who dwells in mount
> > Kailas and who assumes the form of Vishnu".

> > "sá yád dhruvā́ṃ díśam ánu vyácalad víṣṇur
> > bhūtvā́nuvyàcalad virā́jam annādī́ṃ kṛtvā́.

> > "He (Vrata/Shiva), then he went away to
> > the steadfast region called as Vishnu.

The meaning of this narration is that he is the Shiv, Shiva by the big bang get the expansion and touches all the region and known as Vishnu, by his nature of inflation he is called as Space.

Shiv is the formless Brahm while Narayana is formless in manifested form.

In the Yajur Veda, Taittiriya Aranyaka (10-13-1), Narayana suktam, Lord Narayana is mentioned as the Supreme Being. The first verse of Narayana Suktam mentions the words "paramam padam", which literally means "highest post" and may be understood as the "supreme abode for all souls". This is also known as Param Dhama, Paramapadam, or Vaikuntha.

Rig Veda 1:22:20a also mentions the same "Paramam Padam". This special status is not given to any deity in the Vedas apart from Lord Vishnu / Narayana (One who is resting on water). Narayana is one of the thousand names of Vishnu as mentioned in the Vishnu Sahastranama.

It describes Vishnu as the All-Pervading (Inflation as of space) the essence of all beings, the master of—and beyond—the past, present and future, one who supports, sustains and governs the Universe and originates and develops all elements within. This illustrates the omnipresent abiding characteristic of Vishnu.

Vishnu governs the aspect of preservation and he is the substratum of the universe, so he is called "Preserver of the universe".

Vishnu is the Supreme Deity or Godhead who takes manifest forms or incarnation across various Yugas or eras to save humanity from evil beings, demons or Asuras.

Prajapati Brahma Said:- Hey Narad, Vishnu salute to Sada Shiv-Shakti and then said, Hey excellency who I am and what are my duties?

Shiv Said:- Because of your entelechy enormousness Spacious you will famous by the name of **Vishnu.** You will also get many names that reveal the devotees. And now you should penance which will escalate all desires.

Thereafter from the respiratory path, Shiv thought him all the knowledge of Vedas. Then Vishnu started Penance.

"YASYA NISHVASITAM VEDAH" Meaning: Who exhaled air in the form of Vedas.

That means Lord Vishnu is exploring the knowledge of Vedas as same as air is present everywhere.

And then Vishnu started a tough penance. Sada- Shiv and Shakti then got evanesce (Anthardhyan means disappeared) from there.

Vijay:- Referred to Shiv purana and Shrimad Bhagwat puran, as a result of penance, the first numerous fountain of streams erupted from his body i.e. generated from Vishnu and the space was the abundance of water. He stayed in that water for a long period because of his resting in water, he therefore named as **Narayana**.

Shital:- So you mean that water is the first element of the early universe as per the origin of elements?

Vijay:- Absolutely not, When Lord Vishnu was in penance for the long span, during that period universe goes through many changes. As we will take a further pace of discussion you will find that before the origin of a compound like water many essential elements was formed in the early epoch of the universe. Now just for your understanding initially in the very early epoch of the cosmos after the big bang the universe has only Hydrogen and Helium.

Water was originated very late after the reaction of Hydrogen and Oxygen in the little late phase of chronology of early Universe.

This is the process of beginning of the universe with the **cataclysm** of modern big-bang theory during the 10^{-43} seconds.

Therefore simultaneously all the essential five elements have produced from that Space and thereafter the twenty-four natural elements have been produced. Then he elide all the essential elements and gone for long Sleep (in Yog-Nidra) in that divine water.

These elements are having mass produced from Vishnu (Space) are basically the Shiv. That's why it is said "Kankar Kankar Shankar" which means Shiva resides in every stone and pebble (even in sub atomic particles). So everything found in nature is Shiv himself in a different form of Matter.

* * *

Naradji asked:- Hey Brahma ji, what are those five elements called, I want to know the quintessence of the secret knowledge?

Prajapati Brahma:- Anu, Tamas, Satva and Rajas and the familiar types of characters are driven from Para-Shakti or Brahm is the transforming character as Trinity. These particles smash together to form protons and neutrons. These whole particles smash together called as Atom (परमाणु) and combined form of Matter (कणों) or (Rudra). Space is the (Narayan) and I am the Time (Brahma). All that you see in this world in the united form is SadaShiv the **Brahm**. These were the five fundamental particles from which the multi-verse was evolved. Science of any era may give any names to these particles. But truly it will remain constant as Rudra particles. You will get more clarity on Rudra (Matter) along when take a move with this story.

Vijay:- The entire story of light and heavy elements, fundamental forces and epoch of the universe will be decoded in our further discussion gradually step by step knowing which will be the end of fantasy and beginning of wisdom.

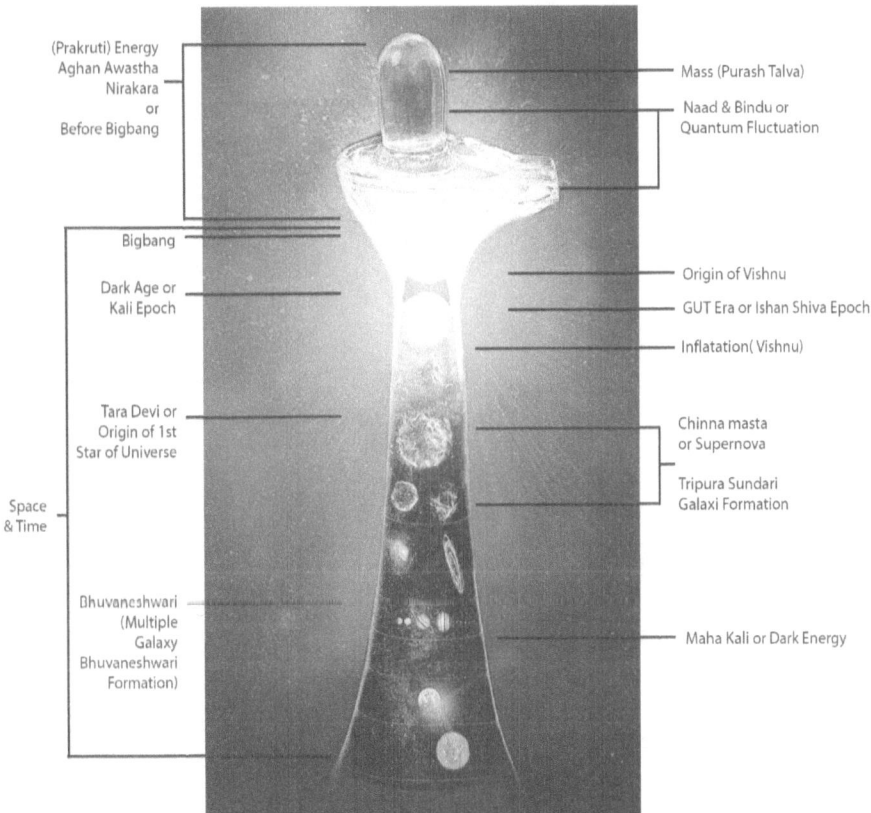

(Prakruti) Energy
Aghan Awastha
Nirakara
or
Before Bigbang

Mass (Purash Talva)

Naad & Bindu or
Quantum Fluctuation

Bigbang

Dark Age or
Kali Epoch

Origin of Vishnu

GUT Era or Ishan Shiva Epoch

Inflatation(Vishnu)

Tara Devi or
Origin of 1st
Star of Universe

Chinna masta
or Supernova

Tripura Sundari
Galaxi Formation

Space
& Time

Bhuvaneshwari
(Multiple
Galaxy
Bhuvaneshwari
Formation)

Maha Kali or Dark Energy

**Timeline of the Cosmos, before and after
the Big Bang (Anhad Naad) – Vedic View**

Chapter 5

Chronology of the Universe and Timeline of Cosmological Epochs

5.1 Brahma Epoch - Beginning of Time or Origin of Brahma (Time) from the Navel of the Lord Vishnu (Space) by the Will of Sada-Shiv

Prajapati Brahma Said:- Hey "Devrishi" When Lord Narayan was resting over the water then a lotus cord was sprouted from his navel (umbilical-cord). That cord is an immeasurable and unimaginably huge universe. That yellowish lotus cord itself is a universe.

Thereafter from the right part of Sada-Shiv created me and poured to me in the lotus cord of Shri Narayan, that's the moment from when and where I began as Time; Cosmic Time is something that can seems to be predicted by seeing, hearing, or feeling through any observational method.

That the "uncaused First Cause" is the Creator who exists outside of the physical creation, he made Time is not eternal, but created. All the process has accomplished by energy.

* * *

Vijay:- This can be understood as:

Uncaused first cause- is Shri Vishnu

The Creator – Lord Sada-Shiva (Who exists outside of the physical creation)

He means Lord Shiva made Time (Brahma) which is not eternal.

Therefore the "uncaused First Cause (Vishnu)" is the Creator (Sada-Shiv) who exists outside of the physical creation. He made Time (Brahma) is not eternal, but created. This phenomenon is accomplished by Energy (Adi-Shakti) for the tasks of creation and she is a manifested, un-manifested, and transcendent divinity.

* * *

Brahma Ji Said:- Hey Narad, this is the conte of my birth over that divine lotus, I have four heads and four arms, my memory was very week due to the effect of Sada-Shiv and I confirm that, Lotus to be my filiation (Janak).

Who am I, from where and for what I came, who made me, with long-standing dilemma I thought that water is the source of this lotus cord and it is simple to understand by finding the sprouting point of these cords where I may get my source of origin and my parents.

As decided I went in that cord, I was wandering there for many years and does not find the end of the cord. Then I thought again to reach there at the top of that lotus cord and fail to reach. Then I have started pondering and then I have been inspired to penance.

Then after long meditation suddenly Lord with four hands divine Vishnu appeared and then there was a paradox between both of us, who to be the first among both of us in the Universe.

Then suddenly a perpetuity pillar **Jyothirmay Lingum** appeared in front of us.

I Brahma said:- What is this and from where this flame of pillar came.

Narayan Said:- No doubt, it would have taken a paragon of virtue, do not feel viciously vexed, this is something auspicious and might have some great reason for appearing here. This may be making us some paragnosis or want us to realize something that we don't understand.

Brahma Ji said:- Hey Narad, That was something like a suspense, we are in big dilemma and then we mutually decided to find out the reality therefore we pact to go top to bottom to inspect that divine pillar.

I went to find the top, while Vishnu proceed to detect to locate a bottom of that Jyothirmay Lingum, but we failed to find the top and bottom and return to over respected path and finally we both meet at a junction. Then Shri Vishnu looked in and around of that shining **Jyothirmay Lingum** (Pillar) and we simultaneously said what this object is.

Then we noticed and pondered, started to salute that Divine Pier Jyothirmay Lingum and said O Divine Lord, We don't know who you are, well whoever you are, we repeatedly salute you. Hey Mahashen (name of Shiv) kindly appear and show your actual form. It's impossible to understand you in this way. Therefore we have afflatus and salute that enlighten Pier Jyothirmay Lingum.

We both begun to pray that whoever you are and whatever is your identity O supreme Lord we are pitiable to grant your blessing. While in his praying hundreds of years were passed.

Narad: Oh Father, What actually was that Pier **Jyothirmay Lingum, Please let me know?**

Prajapati Brahma: That Divine Pier Jyothirmay Lingum was scientifically the total Mass (Shiv) accelerated by the Energy (Shakti), was emerged throughout the Space (Vishnu) at constant Time (Brahma) i.e. Me.

So for energy as said in Kunjikastrotram Aim hreem kleem chamunde biche Namha!

Aim is Singularities (Aimkari Sristhi roopay).

That's the Aim, the energy which induces to accelerate a total Mass. That Mass was the Pier Jyothirmay Lingum (Jyotir lingam) emerged and the Space (Narayan) was dazed.

Narayan could not recognize the wonder at that moment. Then I asked Narayan to identify that Jyothirmay Lingum.

By the origin of the Jyothirmay Lingum a sound AUM - AUM (ओ३म्) broke. This was the process of origin of the universe by Maha-Nadh (big bang).

After failed to found the beginning and end of Jyothirmay Lingum, I and lord Narayan together started to pray him, to know his actual form and to reveal his the real existence. Then Shri Vishnu salutes and adores that Jyothirmay Lingum and then the Primordial Sound of a Seer instructed and guided Maha-Vishnu to recognize the Jyothirmay Lingum Mass. This seer is none another then the Lord Shiv Shankar as he is Adi-Yogi.

Space (Vishnu) who is a conceit, now understood the processes of origin of the universe.

* * *

Vijay:- In other words according to modern science we can say that the Space (Vishnu) was observing the Divine Pier Jyothirmay Lingum process of reaction from which the universe began, this phase is called the Planck **epoch** is an era in traditional (non-inflationary) Big Bang cosmology and immediately after this event our known observable universe came into existence.

* * *

Narad: What that Pier Jyothirmay Lingum was made of?

Prajapati Brahma: That Jyothirmay Lingum was the sum of total Mass, whatever you see today in the universe that was present in that enlightened Jyothirmay Lingum. Whole that we see today in Star, Planets, Galaxies, Sun and Moon, Gasses in living and nonliving things of today's universe was found in that primordial Jyothirmay Lingum. Elements and all the scattered matter of today's universe was compactly found in that Jyothirmay Lingum in the form of Plasma.

Lord Vishnu found Dual nature of Radiation and Matter, which is the Energy of that Jyothimay Lingum in form of particle-like and wave-like nature.

The Lord Vishnu (Space) was viewed the Pier, listened the AUM - AUM (ओ३म्) (OM) mystical sound (Sound means Naad in Sanskrit) was the explosion of musical sound elision. That sound was heavily broken.

Shri Vishnu found southern part of that Jyothirmay Lingum was full of Mass (Akaar), He also found the inflation (Ukaar) in all the direction. In the northern region he observed the plasma i.e. helium (Makaar).

Narad:- Hey father, If you are the creator in the form of Akaar in that primordial AUM and Lord Vishnu is the Ukaar as a preserver of the universe in the aspect of divine mystical sound AUM, then how ethereal Jyothirmay Lingum can be Time (Akaar), Space (Ukaar) and Shiv (Makaar) kindly make me more clear?

Prajapati Brahma:- Entire of the cosmos, material, beings and non-beings are made up of three primary characters which are Satguna, Rajoguna and Tamoguna. Though Mass is only associated with these three characters.

The inception Lord Vishnu in the form of Space began with the String (Naad) of Mass (Shiv Lingam) in Singularity (Bindu or Yoni). Thus Vishnu is one character of Shiva and then I was appeared from Vishnu. So I am too the form of Vishnu and Shiv.

* * *

Vijay:- Significantly, Vishnu (Space) is one who is continuously exploring the material or the entities. And Matter (Mahesh) annihilates the cosmos within him at the end of an eon.

Remember Shiva itself is Shakti and Shakti is Shiva. Adi Shakti is also called to be Shiva or Shive.

Scientifically that Jyothirmay Lingum is said to be **The Radiation Era** of **10,000 years.** The first major era in the history of the universe creation in which most of the mass (Shiva) and energy (Shakti) is in the form of radiation -- different wavelengths of light, X rays, radio waves and ultraviolet rays. This energy refies that the primordial fireball.

This energy (Adi Shakti) is the remnant of the primordial fireball (Jyothirmay Lingum) and as the universe expands, the expansion of the universe or Space (Vishnu) the waves of radiation are stretched and diluted until today, they make up the faint glow of microwaves which showering the entire universe.

So the above Story narrated by Shri Prajapati Brahma to Dev Rishi Narad regarding Jyothirmay Lingum was the description of **Radiation Era.**

The Radiation produced after Maha Naad (Big Bang) is called the **Anhal Stamba** in Vedic language, modern theories have endorse the same thing since the last couple of centuries. *Anhal means Radiation and Stamba was the fireball.*

Exact incident are discrived in Vedas, Puranas and Upanishads are based not only observation but with all the evidence and interaction between Supreme and Great Seer's (Ancient Scientists).

The Radiation (The Fireball of Anal Stamba), which as per the Modern theories of Science is the era, which began just after two or three minutes to 300,000 years of Big Bang, with a process known as nucleosynthesis (Origin of Tatva).

This was when helium nuclei (in fact *all* the helium nuclei in the universe) were formed out of protons. So this is the initial start of the *atomic kingdom.*

Protons react with other protons to create Deuterons. The temperature had dropped to about a billion degrees Kelvin. At this temperature deuterium (the heavy hydrogen nucleus, 2H) was stable. This enabled nearly all of the neutrons and one of the seven protons per neutron to turn into deuterium.

The deuterium quickly reacted with some of the remaining neutrons and protons to form the helium nucleus (originally called the alpha particle) consisting of two protons and two neutrons, 4He. This gives a ratio of 3 protons to 1 neutron. About one in 10,000 deuterium nuclei escaped from being turned into helium nuclei and about one in 10,000 of the helium nuclei remained as the atomic mass 3 isotopes, 3He. This is the same ratio that is found in stars today, (except for the heavier elements produced later) which argues in favor of the Big Bang theory.

For the rest of the Radiation Era (a period of 300,000 years) the universe remained fully ionized, and continues to expand. The entire universe was plasma of freely moving protons, neutrons, and electrons, just as one finds in stars today. Plasma is considered the fourth state of matter (after solid, liquid, and gas) is a state of charged ions (atomic nuclei) and electrons; with photons scattering off them and reionize any atoms that had formed. There are no atoms as exist normally - i.e. electrons orbiting a nucleus, because the heat is so high that any electrons are automatically stripped away.

During the Radiation Era therefore the entire universe was like one single gigantic star, thousands of light-years across. During this period the universe was in thermal equilibrium, and opaque to radiation. The radiation was absorbed by all the free electrons whizzing around.

This is the Modern Scientific theory accepted by almost all the penal of Scientists of the present age. Which they have approved based on their several observations, but have they ever thought that Vedas are narrated during the pre-historical time even before the dawn of any civilization has already proved the same science.

Vijay:- The above Scientific Story I shall narrate it exactly the same way and even better then what the modern Scientists has done. My narration will be based on Vedic and Puranic scripture in support of modern science.

The only thing in this world is a belief system. Once you started to believe, you develop a faith and with faith, reality can be known.

Planck Era as said by Modern science is the closest that current physics can get to the absolute beginning of time. At this moment, the universe is thought to be incredibly hot, dense and turbulent, with the very fabric of space and time turned into a roiling morass. All the fundamental forces are currently at work in the universe - gravity, electromagnetism and the so-called strong and weak nuclear forces - are thought to have been unified during this stage into a single "super force".

Whatever the name was given by modern Science that doesn't matter, what counts are that modern science can change based on the observation, these aspects can differ in the observations based according to the Scientists of various categories.

For Einstein, it is relativity, while for others it is based on quantum mechanics, some other claims the universe is not a changing phenomenon.

Their few observations are fitted at some point of cosmology, but modern science is itself a changing mechanism of observation.

Theories may come and go but the Vedic science will remain same endorsing the reality of ancient actual universal Science.

Shital:- Can you explore a better way to understand these scientific phenomena from origin of the universe to its present form.

Vijay:- Exactly, For this, we will discuss the same science narrated by Nandishwar Ji to Sanat Kumar in the summit of sages held on the bank of river Ganga in the holy city Kashi. The Sages understood the radiation or the cosmic microwave background (CMB).

The cosmic microwave background was the leftover of this Jyothirmay Lingum or Anhal Stambha and Jyothirmay Lingum is the radiation era.

Now, I will narrate to you entire theory from epoch to epoch, pace-wise in Vedic way and at last you will realize there is no much difference in modern and Vedic science in fact Vedic science is much more advance. So let's come to the discussion.

5.2 Appearance of Shiv-Shakti in Ardhanarishvara (अर्धनारीश्वर) Is an Androgynous

Describing the majesty of Annal Stamba, **Sutji** the son of Seer **Ved-Vyas** Says:

O Saint's, "Lord Shiva is the manifestation form of almighty in his unmanifested (Brahm) Swaroopa, for this very reason he is known as NISHKAL. Because of his divine form, Shiva is called SAGUNA (God with form). The term SAGUNA is also expressed in another accreditation i.e. SAKAL. Shivalinga is worshipped since it symbolizes the form of Shiva. Lord Shiva is also considered to be NIRGUNA (Arbitrary - without any qualities.)".

Prajapati Brahma says: Narad, I and Shri Vishnu both are scared and still don't regard that who is behind this refulgence. We both have repeatedly saluted and while doing so 100 years have passed.

Vijay:- (Refer to Shiv Purana) The 100 years was the 10,000 years of radiation era according to modern scientific time scale measurement. Then Shri Vishnu said.

Vishnu Said, O Lord reifies us as we are your refugee?

Narad asked Shri Bhahmaji that has someone answered to you and Shri Narayan?

Brahma: Hey Son, Then suddenly soniferous semblance image of **Ardhanarishvar** (Mass-Energy) appeared and that state was manifest as a united form of Shiv-Shakti which is called as the Ardhnarishwar Swaroop (androgynous).

We have also understood the first and foremost secret of the origin of the cosmos by observing the entire Mass in that Pier Jyothirmay Lingum i.e. (Om Tatbamasi) Tatva is the total Mass of cosmos accumulated with entirely Energy intentionally to bring forth the universe.

Simultaneously we are then knower, decoded Gayatri Mantra, Maha-Mrityunjaya, and Panchaakshari Mantra which is the secrets for every being.

The coupled image of Shiv- Shakti was having all the united features of masculine and feminine that ecstatic incarnation was the wonder of the cosmos.

* * *

Vijay:- During this phase, Matter and energy were inseparable and the four fundamental primary forces were united at this moment. The modern theory of GUT (Grand Unified Theory) says this phenomenon exactly what mentioned in the Vedas. So there is no new discovery. By giving new words or vocabulary ancient Vedic facts cannot be exploited.

5.3 Ishaan Epoch - Grand Unified Theory Era (Gut)

At the Planck time, symmetry breaks and gravity becomes a distinct force. The other forces are still unified as the GUT (Grand Unified Theory) force. This is the start of the GUT era (Ishan Era). Here we have the beginning of Quantum theory and classical general relativity.

5.4 Appearance of "Ardhanarishvara" Shiv and Uma (Shakti)

Brahma said:- Hai Devrishi then suddenly that fireball with fusion get scattered with that a divine coupled Shiv and Shakti appeared in front of us.

This is the shape of their body in materialistic form, though they can be associated with matter and free from it by their will. The divine image is said to be a holistic aseity.

Shiv the auspicious one his attribute posses' five head, having three eyes over each forehead and ten arms representing ten different dimensions, his glory cannot be described by words.

Uma Devi is the supreme energy called goddess Durga by Vedas. The goddess and Its Power (Shakti) manifested as Nature, which is the subject of change.

Durga is refulgent and radiant with ardency; she is the power belonging to the Supreme Lord who has manifold manifestations. She is the Power residing in actions.

This is their Sakal Swaroopa (the manifested form) or the Vyakta roop which comes in the state of appearance.

Shri Vishnu Said: O Lord Shiv, Hey Adi-Shakti, It's our pleasure to have a sight (Darshan) of you holy Lord **Ardhanarishvara persona,** now we know by your grace that you're the only true in abstract and in shape.

O lord we acclaim you, acclaim you, acclaim you?

Hey Adi-Shakti (Supreme Energy) you are the **AIMKARI (singularity)** you have fabricated and accouter this whole universe. That's why Sruti says you a Kusmanda (the fourth form of Nav- Durga) who wears the universe; we salute you Oh mother of thuniverse.

Shiv:- Hey Brahman (Brahma the time), Hey Narayan (Space) you both are not apart from me. What I am is you and what you are is I am.

Brahma you are the right part of my body while you Vishnu get inflation from the left of mine. I am very happy with both of you.

Hey Viranchi Brahma, you are my own representation in the form of Cosmic Time, detectability comes to enforce because of your origin, thus you take the responsibility of further creation and start abiogenesis process, arrange the cosmos in an organized manner. Vishnu you will be the extensively most respectable among all creatures and now you have to operate the law and order of the entire universe.

Vishnu Said:- Hey **Maha-Dev,** Hey Mata **Adi-Shakti** you both are adorable and if you are pleased and happy with us, then always keep your gracious hands over our head.

Mahaswara (another name of shiv) Said: I am the creator, operator of universes and also annihilate universe at the end of an eon. I am the mass, I am the energy. We as Shiv-Shakti are inseparable.

Creation, Protection and Annihilation are my affairs for which I evolved as Space, Time and Matter (Vishnu, Brahma and Shankar).We all are one in three modes of universal operation and the same way the Shakti (energy) will associate us in her different form.

Brahma:- Oh Lord as you have divulge the identity that I am the Kaal (the TIME) and the Shri Narayan is Anantha Antariksha (infinite Space) the outer space and the space inside every being, and it's he who is the expand and expander; celestial bodies, planets, stars and galaxies etc. are positioned within you, you also exist beyond the Earth and between celestial bodies (Shiv the aspect of Vishnu) to operate the all Brahmanda (Universes). Who is Rudra (the Matter) and where it is, what his affairs in this universe are?

Shiv Says:- Hey Shre Vishnu, You and Brahman have worshiped me, for the reason, hey Brahman, Super-Matter is supposed to be disembosom from you, to assist both of you to operate the universe in proper order. That Matter will be known by the name of Maha **Rudra** and will be my divine incarnation. While I am the quintessence (Mass) will appear in the form of matter (Rudra) will fill this space. His key responsibility is to help in creation and to annihilate the antimatter even the entire existence whatever it may be at the end of an eon.

What I am will be that Rudra, substantially to characterize the universe I have to diversify in two distinct forms. Thus there is no difference in Shiv and Rudra (Mass and Matter) Basically entire of the universes is the semblance of Lord Shiv.

5.5 The Summon of Rudra's

Thereafter Prajapati Brahma said to Narad:- Mune, after instructing me and Shri Vishnu for our respective duties Shiv and Shikti get evanesce from there. Then I have taken permission from Shri Hari (Vishnu) to start the creation.

Shri Narayan blessed me and he evanesces from there and pervaded in all directions. I have gone in long meditation in concern to start creation. In the process of abiogenesis, I have taken the element water which was already originated from Space (Shri Vishnu) to start the creation in its consequence the cosmic egg formed but It remains as it is accidie the life could not be generated, which is in a steady-state was in a combination of all twenty four elements.

Hey **Narad**, I was very disappointed with that and then I have started a very hard penance up to twelve cosmic years of the universe. As of my adored Shri Vishnu appeared there and I was really in apparition due to his presence.

Lord Vishnu asked – Hey Brahman, I am very appeased with your devotion and worship, tell me how I can assist you and I am able to give you anything?

Brahma Said:- O Lord, Lord Shiva has given me in your hands, while I have tried many attempts to create this cosmic egg blessed by Lord Shiv alive but it remains in the state of caducity. So now you can only bring momentum in it.

Shri Vishnu Said:- All right.

Prajapati Brahma says Hey Narad, that moment I saw Narayan with Infinite hands, infinite Heads, Belly and feats.

सहस्रशीर्षा पुरुषः सहस्राक्षः सहस्रपात् ।
स भूमिं विश्वतो वृत्वात्यतिष्ठद्दशाङ्गुलम् ॥ १ ॥

Sahasra-Shiirssaa Purusha Sahasra-Akssah Sahasra-Paat |
Sa Bhuumim Vishvato Vrtva-Atya[i]-Tisstthad-Dasha-Angulam ||1||

Then with heavily lightning broke with snarl out over there and he himself entered in that cosmic egg.

Hey Narad, I Brahma created the UND that's why it is called Brahamand (Brahma + Und) in fact every atom of Brahmanda is the combined form of Mass, Energy, Space and Time.

Shri Vishnu is the source of life otherwise everything remains in steady-state of inertia. Shri Vishnu is the **Viraj**, he is both, the creator and the creation that is often personified as the secondary creator, Viraj is born from Purusha.

Shri Vishnu entered in that egg, that's why the universe is named as Brahmanda (Brahma + Und, the Brahma within und i.e. egg) and though it is having all the elements i.e. Shiva Tab.

In other words, Mass and Energy which possess Space and Time are called the Universe. This is scientifically the theory of chemical evolution of life. With his strike lightning of Mahan – Vishnu life originated. That is the power of super sole Vishnu as he is the spirit, an ethereal soul thus Vishnu remains as metaphysically in every being.

Shital:- There are some doubts perceived, May I share with you?

Vijay:- Yes tell me, please.

Shital:- As you told in the narration of Brahmaji to Devrishi Narad, He said that Brahma Ji uplifts the element water to form the cosmic egg. Here I have a doubt, how the element water evolved? As per Vedas, what chemical reaction took place in the prehistoric period of the universe that forms water?

Vijay:- You should note down, that all the elements including water was the late generation elements before these elements evolves, universe goes through many fold changes.

This is the epic of epochs one after another, to reach the cosmos at this state. We will discuss all the phases step by step.

Narad:- Hey Taat **(Taat means father)** Now it is very clear to me that how life is generated. Then who are the being generated very first by you and where they settled, I want to know that.

Prajapati Brahma:- I have started to penance for further creation in the process I have generated four children called Sanaka, Sanatana,**Sanandana** and Sanatkumara from my four Heads.

These all are as same as me. They are as my first mind-born creations and sons of the creator-me (TIME). They are made "beings" solely by the power of my mind and thought in **Brahm**, the four Kumaras undertook lifelong vows of celibacy (Brahmacharya) against my wishes.

Narad Said:- Yes father they are my elder brothers and most respectable too. Can you tell what your wish to them to do and why they refused you?

Prajapati Brahma Said:- As they are the grace of Shri Vishnu and Shiv, they emerged from my brain and all four Kumara's are fully indulge in devotion of Parbrahm. They are the knower of Vedas from their birth and always remain adolescents due to the purity of their mind. I asked them to help me in the process of creation which was refused from them.

As there is no place and no planet to make them stay since they are roaming throughout the universe without any desire to preach the true knowledge of Brahm science in Vedas.

To satisfy and explore my internal desires but four Kumara's as Manas Putras are first step of my creation process. Hey Narad our desires aren't always right and we should not always try to fulfil them and the same I realize after my all four son refuse to help me. Thus I decided to indulge further in the processes (Loka Sanskrit word) in further establishment of progeny.

Hence by the inspiration of Shri Hari Vishnu I have started deep penance to praise the *supreme* divinity for several years. *Supreme* divinity as a result of my hard penance appeared in Ardhnarishwar Swaroopa (androgynous form of purusha-prakriti). He was Umaballabh Shiv in front of me who is in form and formless and amorphous.

He knows everything what was there in my heart and said me.

Ardhanarishwar Shiv-Shiva Said:- Hey Brahma, your are very dear to me and you are my isotopic form, I know your trouble, I am here to help you, express what exactly you want.

I Said:- Hey **Mahakaal** you are immanent, nothing is hidden to you while I am a 'Kaal' (Time) taken birth from Vishnu (Space) and

I and Shri Vishnu is very imminent to you. Oh Lord of lords, by your grace I started the work of creation, but unable to do it so perfectly. So Hey Maha-Kaal (the Lord of Time) helps me out in the proceedings of creativity and cognition of creation.

Shiv Said:- you don't worry as I have told you and Narayan earlier that I will appear from you in the form of Rudra and now the time comes for my manifestation in the form of Sagun-Swaropa to help you and Narayan.

Bhrama Says, Hey Narad **Ardhanarishvara** is the auspicious form beyond any perception is a composite androgynous form of Lord Shiv - Shiva i.e Adi Dev and Adhipara-Shakti manifested.

By doing so suddenly, the lord Ardhanarishvara splited into Maha-Rudra (Mahesh) and Adi- Shakti. Because of big roaring sound of Mahesh I called him Rudra.

Adipara Shakti is the main source of creation. For the creation of universe Adi-Shakti separated from Lord Shiv.

Brahma further say's:- Hey Devrishi Narad, Adi-Shakti and Lord Shiv was willingly forced to separated on my urge of creation. Adhi Shakti granted a boon and said.

Adi-Shakti said:- Hey Shiv, I would be campanion of you and will take a human form, once I will Impeccable the cosmic consciousness of universe.

By saying Goddess that Adi-Shakti is transformed into the cosmic energy of the universe.

By instructed me Lord Shiv evanesce and I have followed his instruction.

The divine feminine part was known by the name of Adipara-Shakti. She is the absolute source of reality the primal energy in formless state gone in the steadfast region.

Vijay:-This was the stage of **Phase Transaction,** which means now the pre existing universe Brahm (Shiv and Shakti) has given whole of mass and energy to that newly cosmos evolved in the form of Matter Rudra, the matter being created out of a quantum fluctuation producing equal amounts of particles and antiparticles, so that 'zero becomes +1 and −1'. And this is often used to 'explain' how the universe popped into existence.

Shital:- So Ardhanarishvara represents the synthesis of masculine and feminine energies of the universe (Purusha the Male and Prakriti the female) and illustrates how Energy (Shakti), the feminine principle of God, is inseparable from Mass (Shiva), the male principle of God. The union of these principles is exalted as the root and womb of all creation.

All right, now after the origin of Mass, Energy, Space and Time the baby spot universe is taking the shape by mean of particle i.e. matter, Dark Matter, Dark Energy and basic fundamental forces in its active state.

But I want to know that how these forces are associated with matter and how it affects the universe in Vedic scientific way and what exactly was that **Viraj?**

Vijay:- The Mass and Energy remain united in manifested and un-manifested i.e. in form and formless. Now in further canto of story you will find the Inertia Shiv and Singularity Adi Parashakti manifested in this known universe in different forms specifies the law of mass and energy (as you knew that energy can neither be created nor be destroyed but can be transformed in different forms and same is for Mass which cannot also be created and destroyed but can be conserved).

Before Maha Naad, Nishkal Shiva desire with a pure consciousness (Chit) for the creation, however cosmos was created by the action of Mula Prakrati, Adi Para Shakti or it's by energy called Maya. This Maya is energy which is really the power having capacity of transformation.

So in short Nishkal Shiva transformed to Sakal Shiva, Same as Devi Shiva or Mula Prakrati or Shakti transformed to Maya Devi.

Sakal Shiva or Nishkal Shiva is one with dual phenomena. It is the Shakti Mula Prakrati having potential state of nature or the state of equality (Samyavasta).

Vedanta describe that Mulaprakriti is equality (Samya) of the Gunas, that is all three Satva, Rajsa and Tamsa Guna remains in the state of equilibrium in Mulaprakriti, which has activity (Kartrittva), but no consciousness (Chaitanya).

The connection between the Shiv and Shiva (Adi-Shakti) is one remain unseparated (Avinabhava Sambandha). Brahm does not exist without Mula Prakriti-Shakti or Mula Prakriti without the Brahm, as these are the two side of one coin. However this state is a singularity.

When that almighty Brahm manifest, it is in the form of Pursha and Prakriti which together create, maintain and then discarnate the cosmos.

The Shiv-Shiva is logically forever and everywhere, they are comport, accompany, coalesce, coincide and co-existed never separated.

The Adi- ParaShakti is the Nature (Prakriti) in the former sense is Mulaprakriti, which means that which exists as the root (Mula) substance of things before (Pra), creation (Kriti), and which, in association with Cit, either truly or apparently creates, maintains and destroys the Universe.

हूँ। इस भौतिक जगत्का जो कारण है, उसे 'प्रधान' कहते हैं। उसीको महर्षियोंने अव्यक्त कहा है और वही सूक्ष्म, नित्य एवं सदसत्स्वरूप प्रकृति है। सृष्टिके आदिकालमें केवल ब्रह्म था, जो नित्य, अविनाशी, अजर और अप्रमेय है। उसका दूसरा कोई आधार नहीं है। वह गन्ध, रूप, रस, शब्द और स्पर्शसे रहित है। उसका आदि और अन्त नहीं है। वह सम्पूर्ण जगत्की योनि, तीनों गुणोंका कारण एवं अविनाशी है। उसे आधुनिक नहीं, पुरातन एवं सनातन कहा गया है। वह ज्ञान विज्ञानका विषय नहीं है। प्रलयके पश्चात् उस ब्रह्मसे ही यह सब कुछ व्याप्त था।

The text from Markandeya Purana

Viraj is the name of the primeval being, Purusha, identified with Vishnu and Shiv. When there was inflation of Vishnu he is regarded as Space and when that Space was filled with dust particle he was proclaimed as Viraj, the one who is having everything within him indicates sovereignty, excellence or splendour.

Viraj is the word of Sanskrit language, in which Vi means Vishnu and Raj means Dust, that is the Vishnu in association with dust (Raj) is Viraj.

The highlights of this phase is Adi-Shakti (Uma) transformed as Aditi in upper part of Space and also other forms.

This is how the conserved Shiva got splited in his various forms and his association with four forces.

Narad was showering with the great divine knowledge of Brahmanda from his father Kaal (Time) Shri Brahma Ji.

He is the Lord Shiva as a first Maha-Rudra called as a Shanker who himself transform into everything what is there in this universe.

Shankar in the form of Maha Rudra (In addition to the cosmic microwave background radiation, another important relic, which unravel the truth behind us is that, all of the matter in the universe today is the composition of this primordial matter (Shankar) contains clues to its origin) the one who tint and hybrid himself in immanent form. That's why Saam-Veda and Rudra Shamita of Rig-Veda Says that It is Rudra who imperceptible this space.

Narad asked:- O Father, how was the behaviour of Rudra, how he looks like, what were the qualities he have and **what was the anti-matter as you says and also please specify about Adi Shakti's different forms?**

Kaal Shri Brahma Ji Said:- First listen about the Matter how it given shape to the universe, it is the gist of whole the theories of the cosmos.

In the regards, I asked him to make the universe of its own kind and thus Maha Rudra started to create everything just identical to him.

Now this Maha-Rudra (Matter), was creating a matter in different forms this is the story of "Shiv Tandav strotam" how matter was emerging through the boundaries of Space.

Therefore, this state of the universe was said to be Hari-Har which is purely the combination of matter and space.

SAT SRISHTI TANDAV RACHAITA, NATRAJ RAJ NAMO NAMAH

HEY AADHYE GURU SHANKAR PITA, NATRAJ RAJ NAMO NAMAH.

He is the Rudra embodiment of Tamas, the inertia, the tendency towards dispersion and annihilation but balanced in Satva and Rajas.

By summon of Rudra, I was full of mirth and ecstasies, which eke out whatever required for completing the creation and thus I came out of pessimism.

Description of Ishaan Epoch or GUT Era

Hey Narad, That was the eon (Kalp) called **Vishwaroopa** in which the universe taking its shape in its primary aspect and by the manifestation of Shiva as Rudra **Ishan** and **Saraswati**.

I remember Shiva instruction which he has given me earlier, about the manifestation of Adi Shakti as Saraswati the goddess of wisdom to be my concert. She (Saraswati) is the consciousness.

Saraswati is the aspect of energy (Adi Shakti) in the form of wisdom and delivering the knowledge in all the species. She becomes my consort and her concert effort and action helps me in the processes of creation, if I Prajapathi Brahma is the creator, Devi Saraswati is the energy of creation. It's only her thoughts with the help of which I am proceeding in the creation. Creation of the universe is our concert effort.

The aspect and the morph of Shiv manifested during this epoch was jewel divine deity ornamented by the law of nature to nurture the cosmos. Though, he is unborn but born for the evolution of matter to give the fundamental forces to universe.

He is imminent formless personality in association with the Adi-Shakti (Primordial Energy).

I eulogized Ishaan Shiv after which four divine entities named **Jati, Mundi, Shikhandi** and **Ardhamundi** are generated, these individuals are of great power.

* * *

Vijay:- Modern science says that the universe before this phase have the temperature of 1 x 10^32 degrees Celsius. This hot thick soup was intense and everywhere.

Hence the fundamental forces **Jati, Mundi, Shikhandi** and **Ardhamundi** manifested from Ishan Shiva. All of them blessed me with the power of creation.

Vijay:-These four are the incarnation of lord shiv and are empowered with all the four fundamental forces. These Forces are **gravity, electromagnetism, weak and strong nuclear forces.** The Epoch of **Jati, Mundi, Shikhandi and Ardhamundi** is the era of Grand Unification Epoch, from 10^{-43} seconds to 10^{-36} seconds:

The force of gravity as a Jati (which remain unified with other three forces) separated from the other fundamental force, after the GUT epoch, the Electroweak Era began, the earliest elementary particles, the state of matter began to be created which is known as Particle Era in modern Physical cosmology.

This GUT era will be followed by the Particle in the next phase of the universe, where the fundamental power of **Jati, Mundi, Shikhandi and Ardhamundi** played an essential role.

The Particles Era leads by the formation of Akaas Lingum, Appa Lingam, Uggra Lingum, and Prathivi Lingum are the Panch Mahabhuta Particles).

1. Akaas Lingum,

2. *Vayu* or air

3. *Agni* or fire

4. *Jalam* or water

5. *Prithvi*: the earth.

> These are the four state of matter (same as mentioned in Modern Science) while Akaas is also the Rudra the Maha Vishnu in the form of space was the first entity of the world.

This is just for your understanding, decently Particles Era or Tatva Epoch will be our course of future discussion.

The Question is, WHERE DOES GRAVITY COME FROM?

The Nandishwar Ji (Incarnation of Rudra) explains this to Sanat Kumar, He says.

Jati which literally means massive, Lord Rudra in the form of **Jati** as a mass produce Akarshan (Attraction) and when a body of heavily mass attracts another object of less mass, then the object with light mass falls towards it is called Gurutva (in English the same is called Gravity).

Gravitation is the characteristics of **Rudra Jati**, who established the law of nature in which one object in the universe attract every other object with a force directed along the line of centres for the two objects that is proportional to the product of their masses and inversely proportional to the square of the separation between two objects.

Vijay:- Lord Mahadev Ishan Shiv Shankar is also known as a Jata Shankar, Rudra Jati (Jata) was the Mass transform of Lord Shiv Shankar into **Jata**.

The significance of **Jatadhari Shiv** is beyond the logic but here it is remarkable to understand Lord Shiva's matted hairs possess quintessential massive numerous powers. Just think how Rudra **"Veerbhadra"** appeared when Lord Shiva smash-up one of his matted Hair in to the ground. This Veerbhadra owns the forces of destructive and destroyed entire Army of Prajapathi Daksha and he hack the head of Prajapathi daksha. I will narrate the complete science of this story in our later part of discussion.

For the better understandings, **Rudra Jati** appeared from the matted hairs of Lord Ishan Shiva pervading in all directions and is the master of Gravitational forces which become the basic principle of all the mass wearing objects.

Rudra Jati is the first fundamental force in the form of gravity separated from the Ishan shiva.

Shital:- I agreed, Big Bang theory, on creation of the universe was propagated by the great scientist Einstein in 1931, which stated that matter, energy, time and space were created in an instant by intelligence outside of space and time. This Statement of Einstein should be noted.

Vijay:- You are absolutely right and well said, that the universe was created by intelligence and also outside of space and time.

Shital:- It means that, Sir Einstein said the same what is stated in Shiv Purana and Vedas that the universe is created by the intelligence and that is Lord Shiva but can you clarify what he mean by outside of space and time?

Einstein stated the matter, energy, time and space are created outside of space and time?

Vijay:- As we have already discussed that the creation began from Naad and Bindu i.e. from Singularity, the Singularity which posses Mass and Energy, the Time and Space are the consequence of Jyothirmay Lingum. As we disintegrate the subject and go deeper in dissection all the doubts will clears.

Modern Science explains the same as:-

General relativity proposes a gravitational singularity before this time (although even that may break down due to quantum effects), and it is hypothesized that the four fundamental forces (electromagnetism, weak nuclear force, strong nuclear force and gravity) all have the same strength, and are possibly unified into one fundamental force called the GUT or Grand Unified Theory.

10^-35 seconds is the time during brief history of universe origin, the vacuum energy density that drove inflation is converted into heat. At the end of inflation the expansion rate is so fast that the apparent age of the Universe, there this is called as the Planck time. The same is called as epoch of Ishan Shiva.

The later status in the epoch of **Ishan Shiva** is so-called Grand Unification Era, in which after the separation of **Jati** as a gravity, other three fundamental forces are still unified with lord Ishana and this epoch was called GUT era in modern science, at the end of which the super force of Lord Rudra begins to break and emerged into further three forces Mundi, Shikandi and Ardhmundi.

The constituent of four forces we see today are **Jati, Mundi, Shikandi and Ardhmundi.**

Bramahi Ji said to Naradji that, the second epoch (Kalp) of universe was known as the **Vishwaswaroopa** because of the manifestation of all the basic four fundamental forces which give shape to the universe. **Vishwaswaroopa** (Viahwa means the world or universe and Swaroopa means the form, shape or figure) means the epoch when the universe was figured out.

As it is rightly said, the aftermath of the universe is determined by a struggle between the momentum of expansion and the pull of gravity. Expansion of Space is due to the attribute of the inflation nature of Maha-Vishnu of his inflation attribute and pull of gravity by Rudra Jati.

This can be described from the reference of NASA which says, the fate of the universe is determined by a struggle between the momentum of expansion and the pull of gravity. The rate of expansion of space (Maha Vishnu) is expressed by the Hubble Constant, H_o, while the strength of gravity depends on the density and pressure of the matter (Rudra) in the universe. If the pressure of the matter is low, as is the case with most forms of matter of which we know, then the fate of the universe is governed by *the density. If the density of the universe is less than the "critical density", which is proportional to the square of the Hubble constant, then the universe will expand forever. If the density of the universe is greater than the "critical density", then gravity will eventually win and the universe will collapse back on itself, the so called "Big Crunch". However, the results of the WMAP mission and observations of distant supernova have suggested that the expansion of the universe is actually accelerating, which implies the existence of a form of matter with a strong negative pressure, such as the* **cosmological constant.**

Nandishwara Said:- Hey Sanat kumar, The **Jati** is the master of force of attraction in the form *Gurutva (Gurt + Tatva)* Mass that wears attraction i.e. gravity. Gurutva means Mass with Akarsan means attraction. Jati just after origin started to perform his duty of holding the matter. He has long reaches, an important role in entire universe, he have control trajectory of all materialistic bodies in cosmos.

Mundi as the strong force separated out from Ishan Shiv (Planck epoch). At this phase the universe began to cool and started to expand.

The power of Electromagnetism as a **Shikhandi** will play the major role in creating the Stars from dust (Renu) particles then leads in originating plancts and galaxies because it is **Shikhandi who's force of** Electromagnetism are responsible for giving things strength, shape and hardness.

Manifestation of Matter in the form of Various **Rudra's** during different phases of the universe was possible with the intermolecular force between individual atoms and molecules as a result of electromagnetic forces applied by Rudra Shikhandi. **Rudra Shikhandi** has the ability to attract and repel the charges.

Ardhamundi wears the week force (week force in modern Science) but an important Rudra controls the mental status of beings. The Rudra that drives the Chandrama (satellite or the electron/lepton) is the weak-force.

* * *

Brahma Ji said:- Hey Narad, Rudra Jati, Mundi, Shikhandi and Ardhamundi manifested from Ishan Shiva. All of them blessed me with the power of creation.

* * *

Shital:- How pretty you have narrated about the eon of lord Ishaan Shiva Shankar (Vishwaswaroopa eon). Can you please describe the GUT as per modern science?

Vijay:- Obviously, The GUT Era began after gravity **(Jati)** has become distinct from other three of the four combined fundamental forces. There are classes of theories called **Grand Unified Theories** **(GUT)** that attempt to describe all forces except gravity in a single framework. The leading type is so-called string theories.

Theorists would say that in the GUT Era the gravity force "froze out" of the universe. The GUT Era lasted from 10^{-43} s to 10^{-38} s. Near the end of this era, grand unified theories predict that the universe cooled to the point that the nuclear strong force began to freeze out, leaving three fundamental forces: gravity, the strong force, and the still combined electroweak force.

This "phase transition" released a huge amount of energy (Devi Maha Maya one of the form of UMA Devi), causing space (Narayan, Vishnu) to undergo a rapid *inflation*. In a mere 10^{-36} s, pieces of our universe the size of an atomic nucleus might have grown to the size of our solar system. We will later discuss observations of the universe that seem to require such extreme inflation. Note that this inflation is very large compared to the speed of light, but again, space itself is what is expanding, so it does not have to obey the speed limit of the speed of light.

This is the scientific phenomena of Jati in the universe.

But, Shital do you know, why lord Ishan named his 2nd son the Mundi?

Shital:- Not exactly, but after knowing the named of lord **Jati** emerged out from **Ishan Jata Shankara,** Logically the name of **Mundi** also deeply rooted with functionality and aspect of Rudra Mundi.

Please narrate the meaning and science behind the lord Mundi.

Vijay:- This is in reference of Shiv Puran and Ling Puran, Mundi was from "Mund" means Skull and Skull posses the brain which is the central and regulatory part of the entire body mechanism of all beings.

Rudra Mundi origin from the Mund (from the skull of Ishan shiva) hence got the name Mundi. Mundi is the strong interaction force responsible for holding matter in same way as Skull in vertebrate holds the brain.

In particle physics, the **strong interaction of Rudra Mundi** is the mechanism responsible for the **strong nuclear force** (also called the **strong force** or **nuclear strong force**), and is one of the four known fundamental interactions.

Look how Brahama ji delineated this secret in his Philosophy of Science.

Brahama Ji Said:- Hey Narad, Rudra Mundi emanated from Ishan with the blaze of powerful force and his power become the aspect of the entire existence, he is associated with each and every Anu (subatomic and atomic particles).

Vijay:- The Rudra Mundi manifestation has extremely major role in every aspect of the universe and the key clamping factor in the **Particle Era** of **universal chronology.**

In the view of modern science, at the range of 10^{-15} m (1 femtometer), the strong force is approximately 137 times as strong as electromagnetism, a million times as strong as the weak interaction and 10^{38} times as strong as gravitation.

It is only Rudra Mundi whose strong force holds most ordinary matter at the nuclear level together because it confines quarks into hadrons particles such as the proton and neutron. In addition, the strong force binds neutrons and protons to create atomic nuclei. Most of the mass of a common proton or neutron is the result of the strong force field energy; the individual quarks provide only about 1% of the mass of a proton.

Shital:- Mundi by its force of strong interaction at nuclear level, holds proton and neutron as same way our brain in skull controls the body mechanism.

Vijay:- When we refer to science, the strong interaction (the coercive action of Mundi) is observable at two ranges: on a larger scale (about 1 to 3 fm), it is the force that binds protons and neutrons (nucleons) together to form the nucleus of an atom.

On the smaller scale (less than about 0.8 fm, the radius of a nucleon), it is the force (carried by gluons) that holds quarks together to form protons, neutrons, and other hadrons particles.

In the later context, it is often known as the **color force**. The strong force of "**Mundi**" inherently has such a high strength that hadrons bound by the strong force can produce new massive particles. Thus, if hadrons are struck by high-energy particles, they give rise to new hadrons instead of emitting freely moving radiation (gluons). This property of the strong force (Mundi) is called color confinements; it prevents the free "emission" of the strong force: instead, in practice, jets of massive particles are produced.

In the context of binding protons and neutrons together to form atomic nuclei, the strong interaction is called the nuclear force (or *residual strong force*). In this case, it is the residuum of the strong interaction between the quarks that make up the protons and neutrons. As such, the residual strong interaction obeys a quite different distance-dependent behavior between nucleons, from when it is acting to bind quarks within nucleons. The binding energy that is partly released on the breakup of a nucleus is related to the residual strong force and is harnessed as fission energy in nuclear power and fission-type nuclear weapons.

The strong interaction is mediated by the exchange of mass less particles called gluons that act between quarks, antiquarks, and other gluons. Gluons are thought to interact with quarks and other gluons by way of a type of charge called color charge. Color charge is analogous to electromagnetic charge, but it comes in three types (±red, ±green,

±blue) rather than one, which results in a different type of force, with different rules of behavior. These rules are detailed in the theory of quantum chromodynamics (QCD), which is the theory of quark-gluon interactions.

Shital:- Everything about the Jati the force of Gravity and Mundi's Strong interaction force is elucidated by you, Now I want to know, why lord Ishan Shiva generated Shikhandi, is he have any role in the very beginning of the universe. Why he got the name Shikhandi?

Vijay:- Lord Brahma was blessed by the four fundamental forces which are distinctly the essential for all the aspect of nature, Shikhandi is quintessence force like other three forces from the particle era till today's universe.

Reference: The weak force, or weak interaction, is stronger than gravity, but it is only effective at very short distances. It acts on the subatomic level and plays a crucial role in powering stars and creating elements. It is also responsible for much of the natural radiation present in the universe.

You asked about the meaning of **Shikhandi,** in Sanskrit Shikhandi (Sanskrit: शिखण्डी) literally means lock on the crown of the head. It resembles lord Shiva's head top curly hair which is like a peacock's crest which has a gold shine. Therefore weak force of **Shikhandi** is actually an attractive force that works at an extremely short range of about 0.1 percent of the diameter of a proton, according to *Hyper Physics.*

After that the era called **Electroweak Era** began, which is the epoch of Shikhandi and Ardhamundi.

During this era, the electromagnetic **(Shikhandi)** and nuclear weak forces **(Ardhamundi)** are still combined. The temperature of the universe at this stage is more than 10^{15} K, and there are no ordinary particles yet, just photons and pure energy. We do have a complete theory that can be used to understand the universe at the end of this era. By the time of 10^{-10} s, the temperature cools below 10^{15} K, and finally, the last of the fundamental forces, electromagnetic and nuclear weak forces, become distinct. We have also done particle physics experiments at energies corresponding to a temperature of 10^{15} K, so we can probe the Big Bang conditions experimentally from 10^{-10} s onward.

Nandishwar Said, Hey Sanat Kumar Ji, by the summon of Lord Ishan and his Son's are the principle of Sound and sound comes in realization because Sound started with just a simple *vibration caused by Sakal Shiva* into Bindu after the vibration leads to a Naad. There were no senses to recognize that sound. So by the origin of Ishan Shiva and the four forces collectively established the principle of Sound because they control the frequency of waves.

Because of their power of controlling the frequency of waves the living creature has the organ "ear" set on a definite frequency to Audible.

These evidences are from Rig Veda, Shiva Purana, Linga Purana and Skandha Purana with the facts of modern Astronomy.

* * *

Vijay:-By evanesce of Jyothirmai Lingum, The temperature of the universe started to fall and we can find the reference of evidence in Shiv Puran. Same has been in the theories of modern time, as the temperature of the universe falls to around 3,000 degrees (about the same heat as the surface of the Sun) and its density also continues to fall, ionized hydrogen and helium atoms capture electrons (known as "recombination"), thus neutralizing their electric charge.

But it should be noted the radiation of Jyothirmai Lingum is absolutely not vanished it remain as leftover which is called the **Cosmic microwave background (CMB)** by modern science.

Reference The physics of universe subjected, Recombination/ Decoupling, from 240,000 to 300,000 years: As the temperature of the universe falls to around 3,000 degrees (about the same heat as the surface of the Sun) and its density also continues to fall, ionized hydrogen and helium atoms capture electrons (known as "recombination"), thus neutralizing their electric charge. With the electrons now bound to atoms, the universe finally becomes transparent to light, made the earliest epoch observable. It also releases the photons in the universe which have up till this time been interacting with electrons and protons in an opaque photon-baryon fluid (known as "decoupling"), and these photons (the same ones we see in today's cosmic background radiation) can now travel freely.

5.6 The Tale of Aeon (Kalp) as a Matter of Universal Law

The Origin of Hydrogen's isotope Deuterium, **Helium and Intermolecular force by the Virtue of Lord Sadyojata Shiva.**

Vijay:- This is the science described by **Nandishwar** to the **Sanat Kumar,** He is the son of Prajapati Brahma, as you all knows that Nandishwar Ji is one of the form (incarnation) of Matter, the incarnation of Shiva.

Maharishi Veda Vyasa narrated the essence of *Sath Rudra Samhita* of Siva Purana to Suta Muni (Seer), the Scientist (Veda Vyas) addressed the congregation of Sages commencing the five major incarnations of Bhagavan Siva:

Sonak adi Rishis asked:- Hey Maharishi, you please describe the various incarnations of Sambhu (Lord Shiva) because of which the cosmos evolved.

Sutji Said:-The Secret he disclosed was best known to him only, none other than Lord Shiva can explore that knowledge. On pondering Lord Shiva Nandishwar said.

Nandishwara: Hey Sanatkumar ji, you are the genius son of Prajapathi. Listen very carefully because these all stories are in concern of establishment of cosmic law's.

Though immanent, omnipresence Lord Shiv has taken numerous incarnations in various Kalp and Kalpaantar (epoch). I will narrate few at the best of my Knowledge.

It was the nineteenth Kalpa which was distinguished by the name of **Shwetlohita,** in which Lord Rudra was manifested as **Sadyojata.**

In that Kalpa Prajapathi Brahma Ji was in a meditation, a young guy of White and Red complexion was begotten (Manifested).

By saw him Brahmaji thought and realizes that he is only Lord Shiva now in the form of **Sadyojata**.

Then Brahmaji was repeatedly saluting him, and then from **Sadyojata** four guys of white complexion was manifested.

They all are having divine personalities with all the power of knowledge and creation. They blessed Shri Brahmaji with all the great cosmic powers.

Nandishwara Said:-

सद्योजातं प्रपद्यामि सद्योजाताय वै नमो नमः।
भवे भवे नाति भवे भवस्व मां भवोद्भवाय नमः॥

Sadyojaatam prapadyaami Sadyo jaataayavai Namo Namah,
Bhavey Bhavenaati bhaveswaam bhavo bhadvaaynamah.

Four disciples are manifested from Sadyojata viz. **Sunand, Nandan, Vishvanand** and Upanandan who would always be in association to his mentor Sadyojata. All four disciples prayed to Sadyojata as: Vandeham Salam Kalankarahitam Staanormukham Paschimam.

Lord Sadyojata is the Morphological aspect of all the vital form of breathing air "Prana Vayu" and his Son's are the aspect of **(apana, vyana, samana and udana)**. They are named as **Sunand, Nandan, Vishwanand** and **Upnandan**. They associated with a great cause and pervaded **Brahmloka** over there.

The **Sadyojata** Shiv has blessed Brahmaji with the wisdom and power of creation.

Lord Rudra as a Sadyojata and his son **Sunand** is the master of - life (प्राण) are controlling the aspect of life through the process of inhalation.

Sadyojata is the aspect of **Prana Vayu**, Prana in a Sanskrit word for "life air" or "life force". It is present all over the universe both in macrocosm (space) and microcosm (bodies of living beings).

Prana is a subtle material energy arising from Rajo guna. It means that "Prana" is the aspect which arises from Sadyojata, is having the character of Brahma the power of creation.

Sunanda as a Vayu (Air) works as an interface between gross and subtle body, enabling all the psychophysical functions.

Although Sadyojata and his son are very closely connected, Prana is perceived by the Jiva which is floating in "Prana" in the heart cavity. The moment of happiness (Anand Ke Chhan) is called Sunand.

Vijay:- The first Vayu (gas) or air, originated at the epoch of nucleosynthesis was hydrogen, hence Lord Sadyojata as a Rudra is referred as Prana.

The Sunand is the first Son of Rudra Sadyojata and is the bond between the atoms of that epoch mainly is referred as the master of hydrogen bonding. The Sadyojata is the Prana i.e. hydrogen which makes the bonding with the help of his son Sunand with other elements.

Prana's movement leads to Jiva's identification with the gross body (SB 4.29.71). Mundaka Upanishad (3.1.9):

esho anur atma cetasa veditavya yasmin pranah pancadha samvivesha pranais cittamsattvam otam prajanam yasmin vishuddhe vibhavatyesha atma

Sunand is the controller of atomic soul can be perceived by perfect intelligence as floating in the five kinds of life airs (prana, apana, vyana, samana and udana). When the consciousness (that pervades from the soul through the entire body) is purified from the contamination of the five kinds of material airs, its spiritual influence is exhibited."

Modern Science may not believe about the Prana, but soul (Prana) is the only aspect of existence, weather it is material or alive. Even subatomic particles exist because of Prana.

We know oxygen is the Prana Vayu for human and the other creature on earth. Life on earth is not possible on our planets without oxygen.

Same way the life of elements at early universe was Hydrogen, with hydrogen and hydrogen bonding with other elements the atomic world begins to grow and perceived.

Prana is one but acts in different ways. Lower Pranas control the senses and are under the control of main Prana. All animate and inanimate things are controlled by the Parmatma. There is a karmic account which is the cause of desire and karma of the Jiva (being).

The **Sunand** stabilized all elements by donating the power of protons in respective of its number in the nucleus of an atom like deuterium, Tritium and so on.

Deuterium (symbol **D** or **2H**, also known as **heavy hydrogen**) is one of two stable isotopes of hydrogen. The nucleus of deuterium, called a **deuteron**, contains one proton (Satva) and one neutron (Tamas Guna). The force which work to stabilize the atomic nuclear of **Rudra Sunand** is **Rudra Mundi** who was manifested during the epoch of Ishaan as a strong interaction force.

Isotopes of hydrogen- The nucleus of deuterium, called a deuteron, contains **one** proton and one neutron is **Sunand**, whereas the far more common hydrogen isotope, protium, has no **neutron** in the nucleus.

Lord Sadyojata, the Rudra is having the aspect of protium having one proton and no neutron in the nucleus, while his son Sunand is the master of Deuterium with one proton and one neutron in the nucleus.

THE THREE ISOTOPES OF HYDROGEN

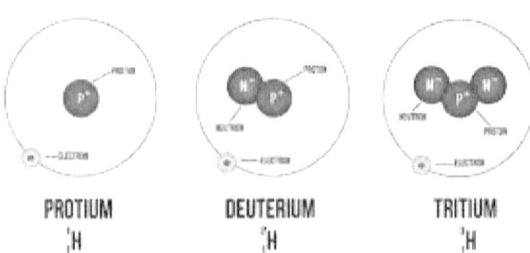

PROTIUM
$_1^1H$

DEUTERIUM
$_1^2H$

TRITIUM
$_1^3H$

Rudra Sadyojata and it's isotopes of Protium, Deuterium and Tritium

These elements were originated from Lord Rudra **Sadyojata** having his own Quality and Merits. He is the son of Sadyojata as a **Sunand** which are the isotopic aspect Deuterium.

The Sadyojata son **Sunand** is the master of various isotopes of elements; He is having the power of protons (Satva guna).

Nandan the Second Son of Sadyojata Rudra is therefore having the character of having equal number of Rajas, Satva and Tamas Guna, as of electrons, protons and neutron.

The Science of the origin of **Nandan** is Helium, atom is created by two Hydrogen's atoms i.e. Sadyojata assembled its two hydrogen atom which are the smallest, basic, stabile, unclosed quantum, wave-particle formation.

Helium is a chemical element with symbol He and atomic number 2. It is a colorless, odorless, tasteless, non-toxic, inert, monatomic **gas**, the first in the noble **gas** group in the periodic table. The "He" boiling and melting points are the lowest among all the elements.

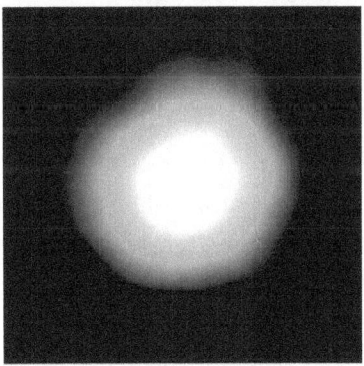

Rudra Sadyojata- The Planetary Nebula of Hydrogen and Helium

This is the result of nucleosynthesis of four elements the Hydrogen, Helium with the trace amount of Lithium and Beryllium.

This is not an automatic phenomena; everything was planned by Lord Shiva and Adi Shakti. It depends over the Time (Brahma) which was required to furnish the cosmos and life.

The Mass of Rudra as a Sadyojata is the Primordial Plasma (Radiation fireball) produced **Sunand** as a representative of **Hydrogen-Bond, Nandan** who was having the power of controlling **Helium** and **Vishwanand** as a intermolecular **bond** is a chemical bond and then the isotopes of **Helium** and **Beryllium** was managed by **Upnandan.**

Sunand is the master of Hydrogen and its bonding (Gandh गन्ध)

He wears the character of bonding between them and also binds the universe, the universe remains only with these four elements for several hundred years.

These four deities are performing various tasks by binding universals atomic particles and their all states of matter.

Sunand the Hydrogen atom by different relationships creates all stabile formations and Nandan Helium atom {alpha particle of nucleus} is the most stabile formation. Atoms structures are bond by energetic part {semi-loop} strong forces of quantum and are very strong bonds but molecules relationships {covalence} are bond by gravity {semi-loop} - weak forces and are bond by electrons.

Vishwanand, the son of Sadhyojat Shiva said to Brahma Ji. O father I will establish the link by bonding Gundh (गन्ध).

The **Vishwanand** is the master of Intermolecular forces; this is the forces between neighbouring molecules, atoms or any other particles. These **Vishwanand** can control attractive or repulsive forces. Attractive intermolecular forces hold substances together and, therefore, these are important to make bulk material. All the molecules have intermolecular forces between them, and some of these forces are weak, and some are strong. **Vishwanand** is the epitome of auspiciousness.

Upnandan:- Upnandan is the son of Rudra Sadyojata, who stabilize the early universe because he is the master of atomic particles.

Sunand, Nandan and Vishwanand all three brothers are under the guidance of Upnandan. All subatomic particles were in voidance and formless.

Thereafter Sunand and Nandan not only established 1st first molecular bond called Helium Hydride (HeH+) but also first chemical compound, to have appeared as the universe cooled after the Anhad Naad (big bang).

All the four brothers together created Brahma-Loka, the planet (abode) of Shri Brahma Ji,

Vijay:- What I have described is the legendary pageant of Lord Shiv and the entire canto is mentioned in Shiva Purana (reference of Rudra-Sanhita, Shiva Puran Page No.420). It was the phase evolved after the Inflation phase of young universe, as it was narrated by Nandishwara that during this era there is white and red complexion.

Shital:- By listening you about the secret of epoch of re-ionisation in your perspective, It seems that this phase of Sunand and Nandan, I mean to say it's the Hydrogen and Helium ions leads to the formation of 1st molecule of early universe.

You also told that Rudra Sunand and Nandan are the morphs of Lord Sadyojata, who together formed 1st bond called Helium Hydride, Isn't it?

Vijay:- Absolutcly right.

Shital:- If it is so, then I have gone through the recently published journal in "The Guardian" with the headline as 'Most ancient type of molecule in universe detected in space'. It also states that Helium hydride is thought to have played starring role in early universe. Can you reveal this?

Vijay:- Yes, This news is fleshed in almost all the media. NASA reported the first type of molecule that ever formed in the universe has been detected in space for the first time, after decades of searching. Scientists discovered its signature in our own galaxy using the world's largest airborne observatory; NASA's Stratospheric Observatory for Infrared Astronomy, or SOFIA.

When the universe was still very young, only a few kinds of atoms existed. Scientists believe that around 100,000 years after the big bang, helium and hydrogen combined to make a molecule called helium hydride for the first time.

SOFIA found modern helium hydride in a planetary nebula, a remnant of what was once a Sun-like star. Located 3,000 light-years away near the constellation Cygnus, this planetary nebula, called NGC 7027, has conditions that allow this mystery molecule to form. The discovery serves as proof that helium hydride can, in fact, exist in space. This confirms a key part of our basic understanding of the chemistry of the early universe and how it evolved over billions of years into the complex chemistry of today. The results are published in this week's issue of "Nature".

It should be noted that Sunand and Nandan are the morph of helium hydride and was the first, primordial molecule. Once cooling began, hydrogen atoms (Sunand) could interact with helium hydride, leading to the creation of molecular hydrogen which in turn is the same Lord **Sadhyojata**.

All four attributes Sunand, Nandan, Vishwananda and Upanandan are menifisted from Rudra Sadhyojata and play the key roles in the formation of molecules.

Shital:- It's so exciting to hear this, helium hydride for the first time in the data. "This brings a long search to a happy ending and eliminates doubts about our understanding of the cardinal chemistry of the early universe.

Vijay:- The Modern science says that this was the phase when Space and Time was there and space (Vishnu) at that juncture had clouds of gaseous.

When the universe was young, before the formation of stars and planets, it was denser, much hotter and filled with a uniform glow from a white-hot fog of hydrogen plasma.

Atomic nuclei (परमाणु नाभिकों) formed by the origin of Sadyojata was the very first pre atomic particles at the beginning of the universe, provides additional clues about events of the early universe and describes about the composition and structure of today's universe.

This is revealed in Vedas, since beginning lord himself has given not only the clue but complete scientific proofs.

The Anhad Naad (big bang) produced a universe was made of entirely of hydrogen and helium. Deuterium is the heavy isotope of hydrogen, was made only at the beginning of the universe; thus, it serves as a particularly important marker.

The white hot fog was the Lord **Sadyojata**, as Sweth (the white color) and Lohita (Reddish) is the radiation of Plasma hydrogen.

The mixture of red and white make a purple color called **Sweta Lohita** is the color of primordial hydrogen in its Plasma State.

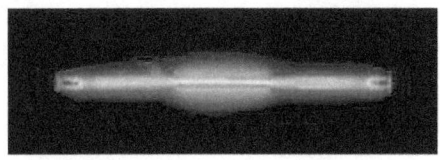

Purple glow in its plasma state

Approximately 73% of the mass of the visible universe is in the form of hydrogen (Sadyojata Rudra). Helium (Primordial Rudra) makes up about 25% of the mass, which we discussed in previous story of **summon of Rudra** and everything else represents only 2%.

Nandishwara said:- The universe before the origin of Lord **Sadyojata** was stagnant and when Brahma Ji found no growth of the universe, he was worried and started meditating on Lord Shiva.

Shital:- Interestingly it's the wonder secrets you are unfolding from Vedas now I am more curious to know about Lord **Sadyojata** and his progeny?

Vijay:- 'Of course' as I told you in Vedic scientific prospective, Sadyojata is the incarnation of Rudra is the aspect of Hydrogen and Helium, which always have key role to perform the growth of the universe.

Here **Mundi** the progeny of Ishan Rudra play the very essential role, it is he who's Strong Nuclear Force binds the nucleus of an atom together. The **Shikhandi's** force i.e. Weak Nuclear Force is responsible for certain kinds of radioactive decay. Same way the son of Sadyojat, the **Primordial Deuterium (Sunand)** is the fused state of supreme Hari-Hara, **Harihara** the state when Vishnu (Hari) infuse with Rudra (Hara) or He is also called **Shankarnarayan** which is the aspect of origin of **Sunand.**

The assembling of Satva with Tamas results the formation of Nucleus and is the Sunand. *Harihara* is also sometimes used as a philosophical term to denote the unity of Vishnu and Shiva as different aspects of the same Ultimate Reality called Brahman.

Thus Lord Shiva appeared in the form of **Sadyojata** having one character of Shri Vishnu as a Satva Guna (the Proton) and one attribute of Brahma Ji as an spinning Rajas Guna (electron). By these two characer there is an element called Hydrogen having one proton (Satva) and one electron (Rajas).

Hydrogen is one aspect of **Rudra Sadyojata** which is the most abundant form element of the entire cosmos. **Hydrogen** is a chemical element with chemical symbol **H** and atomic number 1. With an atomic weight of 1.00794 u, hydrogen is the lightest element on the periodic table. Its monatomic form (H) is the most abundant chemical substance in the Universe, constituting roughly 75% of all baryonic mass.

When we see the atomic structure of hydrogen, in which only one proton and one electron is there, that means at this early phase of the universe, elements are not stable. He is lord Shiva appeared in the form of **Sadyojata** by his donating aspect of neutron (Tamas Guna) begin the phase of nucleosynthesis.

Modern Science describe that approximately 15 billion years ago the universe began as an extremely hot and dense region of radiant energy which is called the Big Bang. Immediately after its formation, it began to expand and cool. The same we have discussed in Vedic phenomena.

The radiant energy produced quark-anti quarks and electron-positrons, and other particle-antiparticle pairs. However, as the particles and antiparticles collided in the high energy gas, they would annihilate back into electromagnetic energy. As the universe expanded the average energy of the radiation became smaller. Particle creation and annihilation continued until the temperature cooled enough so that pair creation became no longer energetically possible.

"Om Bootha Nathaya Vidhmahe,
Bhava Putraya Dheemahe,
Tanno Sastha Prachodayaath"

Upnandan:- Today we find the entire structure of cosmos are made up of the atomic particles and the atomic particles are the result of subatomic particles.

While it is well known in today's world that the fundamental building blocks of molecules are atoms and that atoms in turn are composed of electrons orbiting a nucleus of protons and neutrons.

We have already discussed that in Vedanta and that's the behavior of Satva, Rajas and Tamas Guna. Every being has a set of appropriate balance system of three Guna's fixed by Shiva. When the order of these three Guna changes or disturbed in a being then the matter started to behave as antimatter.

These all Guna's in term of modern science are Quarks, leptons, hadrons, bosons, electrons, protons and neutrons as subatomic particles. The Upnandan is the master of all these subatomic particle so all should fall in a proper order. In simple words the **Upnandan,** controls the world of subatomic particle.

Modern Scientists in fields such as particle physics, astrophysics, and cosmology frequently allude to exotic-sounding particles such as quarks, neutrinos, mesons, gluons, etc.

Sunand is apparently an exotic and esoteric form of Shiva but, in fact, he plays a very mundane role in the world of cosmic existence.

Modern Scientist has to believe that the universe is not running just because of simple theoretical concept. In fact for every incidence of the universe, there is someone who organizes everything. The supreme Shiva and Shakti have given key role to Shri Hari Vishnu Narayana as an organizer. In turn Lord Vishnu appoints the Deities while giving different duties to serve the universe.

If our family, our city, our country and the world cannot run without System then how it is possible to see the cosmic world without the constitution and system.

That's the reason we are having kings, ministers, prime ministers and president in a constitution to run with order.

Simply a family can't be managed without a chief, so the Supreme Lord has appointed a master at every level of micro and macro, living and subtle.

These all incarnation of Shiva or Rudra doesn't mean that it was just the Matter of the universe, they are the Deities who manage the initial universe to grow at it fortune. Rudra's are the multiple representations of Shiva Shankar.

Oxygen is our necessity of life on this earth but there are organisms which can live with different air inhaler, various creature of the universe are surviving at the climatic atmosphere of their world around.

Even on our earth there are Facultative and Aerotolerant Anaerobic organisms can live without oxygen. Thus the Vedanta says, there are bodies with different live force at different place of the cosmos. **Thus Sadyojata is the lord of Creation. He created the celestial bodies like earth and other planets suitable for living creature in universe at young age of cosmic time scale.**

Shital:- Now by listening the science of the growth of universe and the role of Shiv as a Rudras I am more anxious to know such a great Indian scientific literature.

As you told about the recombination epoch which is also referred as cosmic microwave background in which the Universe expanded, however, its **density** and temperature dropped until (after about

380,000 years) the conditions were such that **ions** and **electrons** the Rajas particles could 'recombine' to form atoms (mostly **hydrogen** and **helium** as a) as this recombination was done by Rudra Sadyojata and Perimordial Adi Dev Shankara. This is known as the 'epoch of recombination' or in Vedic language "Sweta-Lohita".

Vijay:- This is from the references of Shiv Purana, page no. 420, The Nandiswar was narrating the secrets of origin of Matter in the form of **Vamdevay.**

Nandishwara was narrating the secret of evolution of matter (Rudras) in various eons (Kalps) and the reason behind their manifestation.

Nandishwara Said:- Hey Brahmaputra, before understanding the Laws of **Sadhyojat** listen the secret of 20th Kalp called Rakta.

In that epoch the complexion of Brahmaji (Time) turns red, your father was meditating on Parmatama Shiva then the Reddish individual manifested over there, His complexion is also red, wearing red ornaments, red fabric and his eyes are also reddish.

By the power of meditation your father understood that he is Lord Shiv manifested in the form of **Vamdeva.**

Brahma Ji Recited:-

वामदेवायनमो ज्येष्ठाय नमः श्रेष्ठाय
नमो रुद्राय नमः कालाय नमः।
कलविकरणाय नमो बलाय नमो
बलविकरणाय नमो बलप्रमथनाय नमः।
सर्वभूतदमनाय नमो मनोन्मनाय नमः॥

The actual description of above Suktam is exactly what modern science says about the early universe radiation.

Vijay:-*Bramaji recited:- I bow to the Noble One (श्रेष्ठाय), the deities of left(वामदेवाय from left portion vast space begotten) who given out the Biggest, the Strongest (ज्येष्ठाय) bonding to the universe,*

Salutation to Rudraya the auspicious Matter and to Time, (The Shiva in the form of Kaalaay)

I bow to the Incomprehensible Radiation (विकिरण or कलविकरणाय), the force is emission of energy as electromagnetic waves or as moving subatomic particles, especially high-energy particles which cause ionization.

You are the source of gamma radiation (विकरणाय) had a significant favourable effect on bond strength to living creatures (सर्वभूतदमनाय) and mark a dent when the adhesive restorative procedure was carried out after radiotherapy.

You are the cause of various forces, and to the Extender of Strength.

I bow to the subduer of all beings, and to the One who kindles the Light.||

बलविकरणाय (Force of Radiation):- The forces generated by radiation pressure are generally too small to be detected under everyday circumstances; however, they do play a crucial role in the world of astronomy and in physics as radiation is the emission or transmission of energy in the form of waves or particles through space or through a material medium. This includes

Electromagnetic radiation:- Radiation like radio waves, microwaves, infrared, visible light, ultraviolet, x-rays, and gamma radiation (γ)

Particle radiation:- Such as alpha radiation (α), beta radiation (β), and neutron radiation.

Acoustic radiation:- Such as ultrasound, sound, and seismic waves (dependent on a physical transmission medium)

Gravitational radiation:- Radiation that takes the form of gravitational waves, or ripples in the curvature of spacetime.

In physics as radiation is the emission or transmission of energy in the form of waves or particles through space or through a material medium. This includes:

- Electromagnetic radiation:- Radiation like radio waves, microwaves, infrared rays, visible light, ultraviolet rays, x-rays, and gamma radiation (γ).

- Particle radiation:- Such as alpha radiation (α), beta radiation (β), and neutron radiation.

- Acoustic radiation:- Such as ultrasound, sound, and seismic waves (dependent on a physical transmission medium).

- Gravitational radiation:- Radiation that takes the form of gravitational waves, or ripples in the curvature of spacetime.

O the auspicious Lord Shiva, I recognize you are the incarnation of Rudra as a Vamdeva, O Lord I am disappointed because of the creation become stagnant.

With all my efforts it is not growing, Hey Mahadev please help me?

Vamdeva Said:- Hey primogenitor, This creation is your image, I will alleviate you from all your sufferings.

Nandishwara Said:- Then from Lord Vamdeva Shiva, the four sons was born to him, they all are very powerful with red complexion too.

They are named as *Viraj, Viwah, Vishoka and Vishwabhavan they all graced your father with the power of creation. Virja creates a complex compounds having a bond of brotherhood Viwah removes the sorrow (Vishoka remover of shoka) is the spirit of world (Vishwabhavan).*

Vijay:- *Rudra Viraja (Viraj) is having the power of creating compound, the compound is the union of two different types of atom (परमाणु),*

Lord Viraja, created a more complex compound by sharing two or more than two different types of atoms, to stabilize the cosmic structure.

When the Raj get merges with the various aspect of space it is called Viraj. Lord Viraj is the united form of Shri Vishnu and Rudra (Space and Matter).

*Before this era (Kalpa) the universe was filled with Atomic (like Hydrogen and Helium) and Subatomic particles. Now **Viraja** the son of*

Rudra Vamdeva is creating and spreading the material throughout the space and interconnected with cosmic structure with the help of bonding brother Viwah.

As by the name of Rudra **Vamdeva,** its clearly understood that the Deva one who is the left part, So energy Adi-Shakti, is the left part of Shiva and in turn Shri Vishnu as a form of Space energy beget from the left part. The Red is the color of the universe at this initial stage.

Rig Veda State, from **Purusha** was born **Viraja**; and from **Viraja** born **Purusha.** Which means with in the first phase of the universe, Purusha was born in the form of inflation of space and in turn again **Tatpurusha** as a Lord Rudra (Matter) has appeared in Space from Brahma, which is the twentieth epoch of manifestation and merging of Matter and Space (Vishnu and Shankara or Hari-Hara). **Tatpurusha** and **Vamdeva** are the two consecutive Vedic eras that will be precisely clear to you in further canto.

Collectively **Viraj** is the gross body of the cosmos personality before this epoch the universe have Subtle Matter, Subatomic and Atomic particles. The **Viraj** is what called to be all the complex matter of universe.

Thus the Vedic words- **Purusha** and **Viraja** are Space and Matter of universe and when the **Energy** (Soul) merges with the matter, Life begins.

Cosmos is the framework of the contents of supreme personality of godhead.

1. Soul in Sanskrit is आत्मन् or Atma (Shiva and the Adi Shakti),
2. Subtle body सूक्ष्मशरीर (The Rudras and his incarnation),
3. Gross body in Sanskrit is विराज् or महत् or स्थूल (Rudra and Narayana or Viraj).

Viraja is what we can see; So Viraj is a **collective behaviour of multi-particle complex systems,** which are proportional to Space.

We found in Puranic and Vedic Narration both Shiva and Shri Vishnu are giving equal importance to each other. Even reverence, honor and worship to each other because nothing perceptible and non-perceptible is beyond them. Cosmos is the matrix of Shiv, Shakti and Narayana along with Brahma, absolutely Mass, Energy, Space and Time.

Viwah the second son of Vamdeva is the bond (बंध), the force which holds and interconnects the matter of the universe together. Which are the bonds common in **compounds (Tatpursa)** are covalent bonds and ionic bonds or it can be a chemical bond. The elements in any **compound** are always present in fixed ratios. Chemical bonds make atoms more stable than they are if non-bonded.

Bond formation involves changes in the electrons (Rajas Guna) on two atoms

This is achieved by one of two methods

- Electron transfer
- Electron sharing

Now we know the modern scientific theories of compound formation by chemical bonding. Look how beautifully same is described in Vedas and Puranas.

Shital:- Its very true but Vivhah is the wedding ceremony, it's a traditions viewed as one of the SAMSKARA, which is a lifelong commitment (Bond) between wife and husband, Is it having any relation with the Lord Shiva's son Vivhah?

Vijay:- Obviously, The **Karma (Duty)** of **Deities** are having same effect on human, demons even on animals, which they have on cosmologically over the inanimate senseless celestial bodies.

The ritual 'Viwah' is a sacred bond of love and affection which binds two different man and women, for whom they may not know before their bonding begins; therefore this law is applicable for every objects of the cosmos, there is a bond of mutual relation between every object of world which bounds the objects in mutual relationship and same way bond between the molecules are called **Viwah.**

Vishoka:- Vamdeva's son Vishoka is a great yogi, the Seer is the symbol of auspiciousness, remover of sorrow and grief of Shri-Brahma Ji, He is the factor of happiness for universe. He is reddish complexion; *it represents unmatched force that is capable of transforming all elements of the cosmos. Uplifts the element of Tejasa, He is the one who engage in destroying Antimatter.*

In the process of creation Prajapati Brahma Ji, suffered a lot and when he failed to create exactly what he wanted, he was in a pain and Maha Yogi Vishoka along with his brother removed all his pains and sorrow.

Vishoka has given stability and clarity of mind to Prajapati Brahma Ji, which is necessary *before* being able to experience the subtler meditations in the state of Samadhi.

Vishoka = state or one who make free from pain, grief, sorrow, or suffering,

Vishwabhavan as Intra-molecular Forces

The force of Vishwabhavan creates the molecule like water by the bonding of two different molecule of Sunand (Hydrogen) and oxygen.

Nandishwara narrated the wonders of secrets to Sanat Kumar Ji, He said by the advent of Vishwabhavan the law of Intra-Molecular forces applied and binds the matter of the universe. These are the forces between the atoms of a molecule or compound. They bind atoms to each other and keep the molecule intact. There are three types of intra-molecular forces as covalent, ionic and metallic bonding based over the Satva, Rajas and Tamas guna (qualities).

When two atoms having similar or very low electro-negativity difference, react together, they form a covalent bond by sharing electrons. Moreover, atoms can gain or lose electrons and form negative or positive charged particles respectively. These particles are called ions. There are electrostatic interactions between the ions. Ionic bonding is the attractive force between these oppositely charged ions. Metals release electrons in their outer shells and these electrons are dispersed between metal cations. Therefore, they are known as a sea of delocalized electrons. The electrostatic interactions between the electrons and cations are called metallic bonding.

Vijay:- Later on after the epoch of Dark Ages this, **the force of Vishwabhavan creates the molecule like water by the bonding of two different molecule of Sunand (Hydrogen) and oxygen. So Vāmadeva** is the Rudra of- Preservation, North, and Water (Jala) Reference from Shiv Purana.

Vijay:- In chemistry, a compound is a substance that results from a combination of two or more different chemical elements, in such a way that the atoms of the different elements are held together by chemical bonds that are difficult to break. These bonds form as a result of the sharing or exchange of electrons among the atoms. The smallest unbreakable unit of a compound is called a molecule (अणु).

A compound (Tatpursa) differs from a mixture, in which bonding among the atoms of the constituent substances does not occur. In some situations, different elements react with each other when they are mixed, forming bonds among the atoms and thereby producing molecules of a compound. In other scenarios, different elements can be mixed and no reaction occurs, so the elements retain their identities. Sometimes, when elements are mixed, the reaction occurs slowly (as when iron is exposed to oxygen); in other cases it takes place rapidly (as when lithium is exposed to oxygen). Sometimes, when an element is exposed to a compound, a reaction occurs in which new compounds are formed (as when pure elemental sodium is immersed in liquid water).

Often, a compound looks and behaves nothing like any of the elements that comprise it. Consider, for example, hydrogen (H) and oxygen (O). Both of these elements are gases at room temperature and normal atmospheric pressure. But when Hydrogen and oxygen react or combine it forms Water (H_2O).

Sanat Kumar:- The Rudras are the aspects of cosmic presence, all the universal law's begun with the summon of Rudra's. Hey Naandishwar Ji, as you told about the matter of universe, now I am eager to know about Tatpursha because I know the lord of lords **Shri hari Vishnu** is the Purusha, is there any relation between Purusha Shri Narayan and Tatpurusha as a **Shiv Shankara**. This cosmic body together with the Supreme Spirit Narayana is known as Anadhi-Pinda (Shiva) or Para-Prakriti (Adi Shakti as primordial energy) combined to become **Tatpurusha**.

Every living being in the world is the yogic association of **Tatpurusha** i.e. the body (Pinda or Tatva) when associated with **Purusha** (Athma) forms a compound and the process is executed by **Rudra Tatpurusha**.

Nandishwara Says:- As you know Lord Shiva and Shri Vishnu are the two aspect of Parbrahm the supreme. Shri Vishnu as a Purusha is the vast Space and vast Tathpurusha is the Matter and have the responsibility to run universe.

That was the twenty first Kalp of the universe which is known as a Peetwasa Kalp when Shri Brahma Ji was in the state of deep meditation he turned as pale complexion. In meditation he was in thought of having a son, who should govern the further universe.

Suddenly a Juvenile luminary individual ornated with yellow surroundings appeared who is having endless arms. By his gesture of meditation your father recognizes him as a Tatpurusha Shiva. Brahma ji saluted him by the recite of a Mahadevi Shankari Gayatri.

ॐ तत्पुरुषाय विद्महे. महादेवाय धीमहि. तन्नो रुद्रः प्रचोदयात्.

Om *Tath Purushaya Vidmahe* Mahadevaya
Dhimahi Tanno Rudra Prachodayat.

Let us invoke the Tatpurusha (the scientific notation is -Tad or Tath means the eminent lord which mean the whole stuff or matter of the cosmos and

Purusha which not merely a male but the whole space or soul spirit) and Immanent Lord (vidvamahe). Let us meditate and focus upon (dheemahi) the supreme Lord (Maha Devaaya). Let us ask Shiv (Rudra) to provide us inspiration and guidance (prachodayaat) in our spiritual journeys.

By this Mahadev Ji was pleased and then numerous Luminary guys (spirits) the Rudras appeared from the Tatpursha Mahadev and they all are the aspects of yogic association of cosmos.

This was the era of multiplication of Purusha in order to perform the various yogic tasks at the different phases of the universe. Therefore Rudra **Tatpuruṣa** performs the role of concealing the universe by his grace and later on controls all the form of Maruds or Vayu (Air).

Shital:- Up to now you have glorify the initial epoch of early universe which break the grand unified theory as a Shiva, the splitting of four fundamental forces from each other by **Rudra Ishan**, with the origin of subatomic, subtler and atomic particle. You briefed about formation of molecules and compounds by the help of Lord Adi-Dev Mahadav by his various incarnations of Rudras and Prakrati Adi-Shakti (energy).

Now I want to know further eras after Anath-Nadha (Big Bang) and science of the origin of this visible cosmos of Stars, Galaxies and Planets?

Vijay:- The universe up to the twenty Kalpa was illuminated by the energy of radiation which was produced as a by-product of **Jyothirmay Lingum** originated from Maha-Naad. Then the Shiv, Shiva (Adi-Shakti) and Narayan decided to have a universe which should be govern by certain Deity's with their key responsibilities.

The **Divine Jyothirmay Lingum,** what we called in modern science is the radiation left by The *Cosmic Microwave Background (CMB)* is the radiating energy of Shiv and Shakti. We have already discussed (*CMB*) by the way of scientific method also by practice of Vedic and Puranic cult.

The entire creation is the result of exchange and transformation of matter and energy, the above discussed epoch's of early universe is same the exchange of matter in the form of Rudras and transformation of Adi-Shakti in the form of energies.

Hence, I only conclude that Supreme Brahm is the divine fundamental cause of everything which is perceptible and even imperceptible in the universe. That is the Singularity, that is Nishkal and that turns into Sakal to become Pursha and then into various forms of Rudras.

Vijay:- Now I will tell you some more Vedic Secrets before coming more in the history of Rudras. How the Panch-Mahabhutas, the five state of matter came in existance in the early universe.

Narad:- Hey father, So these are the Rudra's forces they helps you in the process of creation.

Brahma:- In the process of creation of the universe Shiv-Shakti and Narayan appeared in the form of **Panch Mahabhuta**.

Now listen how that matter in the form of Rudra's came in existence, those Panch – Mahabhutas are (Sky, Water, Air, Fire and Earth).

The vibration caused by Shiva (Mass) in Singularities (Adi Parashakti) results lord Vishnu Inflation and also **Aditi** (Dark energy) just after Big Bang. Lord Vishnu as a *Pratham Purush* pervaded in all direction, so that was the origin of first tatwa as a Space or Sky from Shiv and Shakti. Thus he was called as the Rudra in the form of Space or Sky.

Thereafter Vishnu went into long penance, by the effect of Goddess Devi "Yog-Maya" lord Vishnu went to 'Yog-Nidra' by the effect of which Space become constant.

Narad:- O father so it is Lord shiva in the form of Vishnu the first Rudra as a Space (**Akaas**). Now I am very eager to know about other form of Rudras who rules this cosmos.

Brahma says:- Then during the cosmic sleep of Vishnu (Space) enormous streams of water origin from his body. This is also the Rudra (Matter) in the form of water as a **Bhava Lingam.**

Thereafter Rudra in the form of Ugra (Air) appeared which is called to be the **Ugra Lingam.**

Irrespectively then, Rudra in the form of Prathivi (later on by this material planets like earth was formed) was manifested called as **Prathivi Lingam.**

Shital:- That means matter was appeared in the form of sky, water, air, fire and earth. So I am remembering the anecdote when we were on the pilgrimage of Sri-Sailam. You may also remember within the Maiikaarjuna temple campus, there were eight Shiv-Linga and they are names are Akaas Lingum, Appa Lingam, Uggra Lingum, Prathivi Lingum and Rudra Lingum. This shows that ancient Vedas and the holy religious places are full of science.

Vijay:- Obviously, Here Brahma and Vedas want to define, Matter in the form of "Prathivi" should not be taken simply as a earth, in fact "Prathivi" is one of the state of matter called Solid state. Four state of matter are: Water as a liquid (JAL- Taral), Air as a gas (Vayu- which can flu), Fire as Plasma and Prathivi or earth as a Solid) and fifth Space as pre-plasma soup.

Vedanta descrived the principle of Panch Mahabhuta as: (पञ्च महाभूत).

अर्थात,

आकाश में एक गुण=> शब्द (Sound is the principle of Space)

वायु में दो गुण => शब्द + स्पर्श (Sound and Touch or Perceive or Feel is a principle of Air)

अग्नि में तीन गुण => शब्द + स्पर्श + रूप (Sound, Touch and Form or Shape is the principle of Fire)

जल में चार गुण => शब्द + स्पर्श + रूप + रस (Sound, Touch, Form and Taste is the principle of water)

पृथ्वी में पांच गुण => शब्द + स्पर्श + रूप + रस + गन्ध (Sound, Touch, Form, Taste and Odour is the principle of Earth).

These are called **Pancha Bhuta** (*The five phage of matter or the state of matter*). These are **1. Akasa** *means* **Space 2. Vayu** *means* **Air 3. Agni** *means* **Fire 4.** Jalam *means* **Water 5. Bhumi or Prithvi means** *the Earth*. In the initial stages, they are in subtle form, which means they are in the phage of subtler. When these are not even visible are at the stage of subtle for what today's scientists called electrons, protons, neutrons or positron.

Scientists may give any name by their work done in the study of these subtle and gross elements. But in reality, details are given in Vedas and Puranas by the association with their respective action or duties (Karma).

The Vedas elaborates 'Ashta Murthys' (eight forms) of Lord Shiva. Sarva, Bhava, Rudra, Ugra, Bheema, Pasupathi, Mahadeva, Eashana are the eight forms (Murthys) of Shiva.

It was very well narrated by Lord Brahma to Narad, simply to understand the state of matter and how these states came into origin at the beginning of the universe.

Vedic science elaborates the Adhistanas for these eight forms, which are Sarva for earth, Bhava for water, Rudra for fire, Ugra for wind, Bheema for Space, Pasupathi for Yajmana, Mahadeva for Moon and Eashana for Sun.

Now listen how this Panch Mahabhuta has driven out from the creation of five subtle elements called **Sukshma Bhuta Srushti**.

Naradji asked, Hey father so you mean there are some other abecedarian micro-elements from which this creation are made of?

Brahama asked:- Basically like Satva Guna, Rajas Guna and Tamas Guna are the three characters of beings, same way sab atomic particle are the basic building blocks of cosmic web.

Vijay:- Sattva Guna, Rajas Guna and Tamas Guna are also the phenomena of us given by trinity and reflected at a vast in all creatures, same way these sub-atomic particles are enveloped in every resting and even the body in kinetic momentum.

Tamo Guna is the character of Rudra Shankar as a matter which is originated by auspicious Shiv from Bhrama.

Substantially all the subatomic particles are the real transformation of mass (Shiv), Shiv is the auspicious one, devoid of any character in his clandestine, formless state. All the Gunas (Characters) are his own image when he manifested in certain form.

On the other hand Antimatter is having the same Characters but behaves opposite; take Satva-Gunas which is a positive charged but for antimatter this character will be as a negative charged of Tamo Guna, Rajas Guna for Anti-Matter will be the opposite of these characters and same will be for the Tamo Guna.

The evidence from *Shrimad'devibhagwat Puran, in which Lord shiv reciting to Adi-Shakti,*

Shiv said: If Shri Brahma, shri Vishnu have taken birth from you, then am I, Shankar, who was born after them and perform Tamoguni Leela (divine play).

This Tamo character is the outer character of lord Shankara but that's the inside actual guna of Bhrama who exhibit Satva from appearance and Rajas for action.

We are one but manifests as three distinct deities-Brahma, Vishnu and Rudra. We three are the expression of the three natural (**Prakrathi**) qualities respectively- Satva, Rajas and Tamas. And Prakrathi in turn is Adi-Shakti the Durga.

5.7 Pre-Atomic Age or Anu-Poorva Sthiti

BrahmaJi Said:- I will enunciate you the secrete that was the eocene truth of this world which is wrapped by Devi Maha-Maya.

Hey Son, that phase of the universe was called **Anu- Poorva sthiti (Pre- atomic stage).**

That is lord Shiva who exist in the form of **Mass** (Tat) produce a string (Nadh) in **BINDU** singularity (Adi-ParaShakti, the energy which exist before the time) which emerges in the form of Subtle element (Sat) means **Space** called Vishnu and as a Supreme being he is Parmathama, Pratham-Purash or Narayana. Lord Vishnu is imperium who Pervades throughout the boundaries of steady fast region and is rich in Satva guna (Proton name given by modern science for his that character).

Thus lord Vishnu is having all the positivity (so his Satva guna are as in Proton of cosmic web) that's why the behaviour of proton particles are positive but in reality he is devoid of any guna. It was his Yog-Maya which attract all the negative charged particles by the energy.

Lord Shiva manifested himself in the form of Rudra with the aspect of Tamo-Guna (Neutral Particle) hence he is not affected by other particles and their actions. He remains devoid of any illusion and his characters are (as represented in neutrons) present inside the Kendra (Nucleus).

These two (Matter and Space) together are called Hari-Har (Hari is Lard Vishnu and Har is Shiva). Therefore even the smallest atom of Vayu (Hydrogen gas) of universe is composed of these two, only Proton and Neutron.

The atoms mass is in nucleus in the center of an atom, it is the dense area of atom composed of nucleons. These nucleons are nothing but protons and neutrons in Vedic language are the attributes of Hari-Har.

Proceeded in the process of creation Lord Shiva begotten me from Vishnu while I enrich all the particles with negative charged Rajas Guna because of that particles behave as negative ions (Modern scientific name as electrons) and it is Rajo-Guna the subatomic particle revolves in the outer orbits around the Proton and neutron of Anu (atom).

So these are our characteristics, which are present in every particle of this cosmos by mean of Satva, Rajas and Tamas assign to operate and sustain the whole creation.

Devi Adi-Shakti plays the major role to operate the world from **Anu-Poorva sthiti** to cosmic in its gigantic morph by means of different form of energies like Maha-Maya, Yog-Maya and Aditi etc.

Lord Shiva manifests himself in all the living creatures as well as imperceptible things in the form of natural characters associated with five subtle basic elements) Ahankara (ego), Shabda (Sound) Sparsh (touch), Roop (appearance), Rasa (taste) and Gandha (smell).

Lord Shiva associated as a mass (Neutron the modern scientific name) with an energy to form Nucleus. Narayan generated **Parmanu** in his cosmic sleep. It is simply to understand by the accumulation of two or more Parmanu forms 'Anu'. Two or more Anu (Atom) accumulates to form Molecule and ultimately appears as a matter of the universe in due course of time.

Vijay:- Shital, you will find nowadays there are few religious peoples who claims that subatomic particles like electrons, protons and Neutrons are Brahma, and Vishnu and Mahesh.

I respect the knowledge and work done by them with devotion but they should go through the deep in Vedas and Puranas to enrich their knowledge. The trinity and Adi- Shakti are beyond our perceptions.

There is nothing apart from Mass and Energy (Shiva and Shakti). Brahma, Shri Vishnu and Maha - Rudra all are one (**Ako Brahm Dujo Na Asti**) as Sada Shiv remain as the impersonate derivatives in different form of their **Triguni Maya.**

Sub-Atomic particles are nothing but the characteristic of **Tirguni Maya** (Satva, Rajas, and Tamas Gunna) of trinity serving in the behaviour of being and non-beings by the energy **Yog-Maya.** This Maya Devi is the Nature (Prakati) of Brahm (Krishna is the Swaroopa (form) of true Brahm). Same theory of **Anu- Poorva sthiti (Pre- atomic stage)** is given in Shrimad Bhagwatam.

Yog-Maya is the energy which binds the Subatomic particles like electron, proton and neutrons by various bonds.

Same knowledge has been narrated by Lord Shri-Krishna to Arjuna in the battle field of Kurushetra around 5000 years back. Which is given in Chapter 7th of Shrimad Bhagwat **Gitaji?**

Shital:- Shall you deliver that scientific knowledge of Shri-Krishna for the understanding of today's generation.

Vijay:- Once Same has been asked by Tirupathi Bannot (Hyderabad) same I will narrate you in brief because so many scholars done the nice treasure work on this, I will try to add small drop in their research. *Lord Krishna preaches Arjuna about Bhakti yoga, Dhyana yoga & also Jnaana yoga. In this process, Arjuna becomes illuminated.*

SLOKA

Gyanam te 'aham sa-vigynanam
idam vaksyamy asesatah
yaj gyatava neha bhuyo 'nyaj
jnatavyam avasisyate

SYNONYMS

Gyānam— knowledge; te—to you; aham—I; sa—with; *vigynanam* — scientific knowledge; idam—this; vakṣyāmi—shall explain; aśeṣataḥ— nothing left; yat—which; *gyatava* — by knowing; na—not; iha—in this world; bhūyaḥ—further; anyat—anything more; jñātavyam—knowable; avaśiṣyate—remains to be known.

Shri Krishna Said:- O Arjun, Now I shall unfold you the wisdom of myself (i.e Brahm) in formless (Nirguna) aspect along with the scientific knowledge of supreme head in qualified aspect. By knowing which nothing else remains yet to be known in this cosmos.

Krishna is unfolding here the secret, that he is absolute formless aspect as he is there forever with time and before the time. The universe may goes to absolute zero like before the big bang. That "Brahm" or "Parbrahm" is an aspect in formless and he is absolutely qualified form throughout the adinfinitum of cosmos.

SLOKA

manusyanam sahasresu

kascid yatati siddhaye

yatatam api siddhanam

kascin mam vetti tattvatah

SYNONYMS

manuṣyāṇām—of human; sahasreṣu—out of many thousands; kaścit—someone; yatati—endeavors; siddhaye—for perfection; yatatām—of those so endeavoring; api—indeed; siddhānām—of those who have achieved perfection; kaścit—someone; mām—Me; vetti—does know; tattvataḥ—in fact.

TRANSLATION

Among thousands of human, one may endeavor for perfection, and of those who have achieved perfection, hardly one knows to me in truth (nuance).

Thousands of being tries to know me by their endeavor to know me perfectly but one among that thousands attach me with his devotion to know me in science.

SLOKA

Bhumir apo 'nalo vayuh

kham mano buddhir eva ca

ahankara itiyam me

bhinna prakrtir astadha

SYNONYMS

Bhumiḥ—earth the State of Solid; āpaḥ—water, State of liquid; analaḥ—fire; the state of Plasma; vāyuḥ—air i.e. Gaseous state; kham—Space; manaḥ—mind; buddhiḥ—intelligence; eva—certainly; ca—and; ahaṅkāraḥ—false ego; iti—thus; iyam—all these; me—My; bhinnā—separated; prakṛtiḥ—energies; aṣṭadhā—total eight.

TRANSLATION

Earth, water, fire, air (Four states of matter) Space, mind, intelligence and false ego-altogether these eight comprise my separated material nature energies.

These are matter and energies associated with me, by virtue of which the whole cosmos operates. Though these indeed are my lower nature by which all beings are in delusion. But when one comes in the state of controlling these material and their senses he will come to me.

SLOKA

Apareyam itas tv anyam
prakrtim viddhi me param
jiva-bhutam maha-baho
yayedam dharyate jagat

SYNONYMS

aparā—inferior; iyam—this; itaḥ—besides this; tu—but; anyām—another; prakṛtim—energy; viddhi—just try to understand; me—My; parām—superior; jīva-bhūtām—the living entities; mahā-bāho—O mighty-armed one; yayā—by whom; idam—this; dhāryate—being utilized or exploited; jagat—the material world.

TRANSLATION

Besides this inferior nature, O mighty-armed Arjuna, there is a superior energy of mine, by which are all living entities who are struggling with material nature and are sustaining the universe.

Shri Krishna Says O Mighty Arjun, which is a superior energy of mine which means the energy in the form of Yog-Maya which is associated with Space, is to sustaining of the universe.

Shital:- You have provided considerable empirical evidence in the support of all my arguments. Unbelievable, you have narrated so scientifically, now I understand and acknowledge that everything we see today in this cosmos is the only one, the true Brahm. Who is separated in different entities in the form of object and subject?

5.8 The Wrapping Kali or Epoch of Dark Age and Appearance of Rudra Aghore

Vijay:- Prajapati Brahma was worried and do not found the way to come out of such miserable Phase.

He was seated on the throne and relaxed, diverting his attention to destiny of Supreme Lord; he thought that before something unusual will happen and to escalate from such eccentric state, Brahma Ji set his mind on the Deva Adi Dev Mahadeva lord Shiva and Adi Shakti, started to meditate on Divine Braham Swaroopa of Shiv and Shakti.

Meanwhile the Dark phase was already started; everything went out of glow. The universe reached dimmer and ultimately went into the great stage of darkness. The Goddess **Kali** has taken everything in her raging of rolls.

Commonly it is believe that the mother **Kali** is terrible and is the symbol of wrath but the cosmic reality is beyond this, she is the universal truth same is being described by modern scientific theories as a **Dark Age.**

Dark Age (or Dark Era), from 300,000 to 150 million years: Is the age of Kali which was manifested after the origin of Panch Maha Bhuta and exist until the manifestation of Tara Devi (The First Star). Tara Devi is the second among Ten Wisdoms (Dus Mahavidhya).

The Modern Science says, it as a Dark Age whiles it is the effect of mother **Kali** as per Vedanta. Let's take few versions, first as per modern science the period after the formation of a first **Anu** (Atom) and before the origin of first Star (Tara Devi) of the universe is referred as the Dark Age.

Vijay:- The **Goddess Kali** sprawl's and spread herself evenly and constantly throughout the universe, not only in space but also in time in other words, her effect is not diluted as the universe expands. The even distribution means that dark energy does not have any local gravitational effects on her, but rather she exerts she exert a global effect on the universe as a whole. This leads to a repulsive force, which tends to accelerate the expansion of the universe.

Science and Vedas both claims the same that the dark energy does not have any local gravitational effects, which means the Adi Shakti in her Kali Swaroopa (Dark attribute) have no interaction and association by any mean with the master of Gravity i.e. with Rudra Jati. Evidently, Dark Energy in her Kali aspect posses a vast repulsive force and with the power of this force, the Space as a Lord Vishnu expand more and more.

Although the photons that exist was the result of Radiation era or the era of recombination which was remain over there, the universe at this time is literally dark, with no evidence of light, because earlier to this phase all the cosmic light is due to the radiation of **Jyothirmay Lingum**. As Jyothirmay Lingum evanesced the cosmic temperature falls gradually.

By the arrival of Rudras only very diffuse matter remaining, activity in the universe has tailed off dramatically; with very low energy levels and very large time scales. Little marks happens during this period, and the universe is dominated by mysterious "dark matter" (the dark matter is Aghore Shiva as a Rudra) therefore this is the age of **Kali**.

The Antimatter at this phase was very arrogant and destroying the matrix of cosmos and power of Kali dominated the universe and **Aghora Shiva** annihilates all the antimatter.

During these "Cosmic Dark Ages," The Goddess mother **Kali** and **Kaal – Maha Rudra (Aghore Shiva)** evoke **Sunand** the neutral hydrogen gas as to dominate the universe.

Then Rudra **Sadyojata Said:-** Oh **Aghora**, I am your **adjuvant comrade** and will help you in this creation. (The interaction is not by mean of words, it's by the action).

Now Sadyojata evocated all his four sons who are the substratum of cosmos, the four principle of Brahmanda and all participated in creation,

Sunand (clouds of primordial hydrogen gas) started to collapse with the force of attraction with the gravity of **Jati** and the matter within this Dark Age of Kali and her energy manifested as a first star of the universe.

The Primordial Adi Shakti Devi Kali with in her womb mingles with the matter of Sunand (clouds of hydrogen gas) by the gravity results in the formation of small clump. This infant star started growing rapidly because the **Jati** by his force of gravity trail the matter from all direction of space. Now there was enough matter in one place that the temperature got high enough for nuclear fusion to begin - providing the engine for stars to glow.

This is the time of manifestation of Tara Devi (The 2nd Deity among Dus Mahavidhya, the very first Star of the universe). This is the history of very early phase of universe when the infant star nuclear reactions emitted ultraviolet radiation, stripping the surrounding hydrogen atoms of their lone electrons, making them ionized.

The Vedic Story goes as when Sanat Kumar Ji asked to Nandishwar about featuring the role of Rudra in the processes of creation.

Shital:- The 1st Star of the universe, Oh that's great to know about Tara Devi the very 1st Star, amazing after knowing the science and Vedic theory, I understand, really the Vedas and Puranas are having deft of knowledge. Please narrate me more about the hidden secret of Tara Devi and this phase of the cosmos.

Vijay:- Absolutely, I will reveal more about Goddess Tara Devi in our journey of cosmos, you have to wait up to the canto of Das Maha-Vidhya's (Ten Wisdoms) description or on the point of Uranography epoch after dark age in the timeline of the universe.

Why I say our journey of cosmos because we all living being and non living are made up of the remnant of stars, so we are the part of universal staff from the time of origin of some very 1st stars of universe. Whole stuff of the universe is having the composition or elements delivered from the 1st Star. Let's take a move with Nandishwar Ji narration.

Sanat Kumar:- Hey Nandiswahar Maharaj, Please narrate the further incarnation of lord Shiva as a Rudra.

Nandishwara Said:- Hey **Sanat Kumar Ji,** before describing the epoch of **Aghora.** I want to enlighten you more about the incarnation of Lord Rudra in the form of "**Tatpursa**" Who is the united form of Vishnu and Shiva (Mass and Space).

"**Tat**" means Shiva as a Rudra and Pursa means Narayan, both are conjugated, compeer and originated as a **Tatpursa. Tatpurusha** is the Eastward incarnation of Maha Siva being of yellow complexion. Invocation to Tatpurusha states: *Tat Purushaya vidmahe Maha Devaaya dhimahi tanno Rudrah Prachodayaath.* Salutation to this aspect of Siva is: The appearance of Tatpursa as yellow color.

Vijay:- Nandishwar said, that Kalp (Kalp means eon) was known as a Peetvasa and son of Prajapathi Brahma Ji was enwrapped a yellow fabrics (Fabric does not mean merely clothes it was the fog of gases at this phase in the chronology of universe).

This is the Kalp (eon) after the 20th phase as per Vedic, after the manifestation of Tathpursa as a epoch of subatomic particles assembling by the bonding called Covalent Bond (सहसंयोजक बंधन).

The Rudra Tatpursa is literally the Compound having numerous arms (The dimensions) scattered every where in the cosmos.

A **covalent bond** (सहसंयोजक बंधन), also called a molecular **bond**, is a chemical **bond** that involves the sharing of electron pairs between atoms. These electron pairs are known as shared pairs or **bonding** pairs, and the stable balance of attractive and repulsive forces between atoms, when they share electrons, is known as **covalent bonding**.

When creation becomes stagnant and the new aeon is about to start Brahma ji was worried about how to govern it further. It was about to start the twentieth Kalp called as 'Rakta' which is red in color.

During that Time a reddish color individual was appeared in front of Brahma Ji, his attribute is Red color. His limbs are red, eyes are red, and his aura was also red.

By saw him, Brahma understood that he is Lord Shiva. And Brahma equalizes him with Shiva given the name Vamdeva.

5.9 The Emanation of Aghore during Dark Epoch of Kali or the Beginning of Aghora Epoch or Dark Matter Era

We have already discussed those eras in detail. Now I will precisely tell you what Nandishwar Ji was narrated to Brahmaputra Shri Sanat Kumar Ji about the epoch (Kalp) called Shiv in which further creation preceded after the summon of Rudra Aghora.

Coming to Lord Aghora, It's the Dark Matter and modern science say's that; the rest of the universe appears to be made of a mysterious, invisible substance called **dark matter** (25 percent) and a force that repels gravity known as **dark energy** (70 percent).

Unlike normal matter, Dark Matter which is the cause of Aghora Shiva and it does not interact with the electromagnetic force as it is the character of Rudra in the form of **Shikhandi**.

The Rudra in this does not absorb, reflect or emit light, making it extremely hard to spot, which is the Character of Lord Aghora Shiva because he mostly rests in a gesture of meditation (*Dhyana Mudra*). There is no one in the universe, which can deviate or Shake him from this state. There is no effect of things over him.

He himself wakeup whenever there is exigency to furnish the welfare of the cosmos. We will see the Nandishwara narration described in Shiv Purana which say's Dark Matter Agora has a super symmetry and extra dimensions.

Nandishwar Ji Said:- Hey Brahamaputra, I told you about the four incarnation of Rudra the fifth and the foremost is the incarnation of Shiv as a Rudra Aghora. This Kalp is known by the name of 'Shiv' the auspicious kalp.

When primeval fireball (Jyothirmay Lingum or Anhal Stambha) had become too cool to produce new light, No object is found to be there to shine. As a result, the universe entered in a period known as the dark age of **Goddess Kali**, the static universe was seems to be *Ekavarna, again d*arken phase for Divine Sahastra Versh (thousand years).

Your father *Brahma grieved again for genesis,* was very much worried, and started to meditate upon Anadi Shiv-Shakti. Then after certain period of time a Great Lustrous young guy with black complexion was appeared in front of him. *He wore black; he had a radiant black crown and a black sacred all around his jet body.* Though he is black in appearance but he is the eponym of luminous with great radiation. While his Tinge, Lustre, Shade, Shine, and Tincture are black.

After sawing the Divine, Great Lord, Lustrous, Supernatural, Uncanny, Weird and Eerie, Brahma Ji effaced all his worries and prayed to him with jubilant.

अघोरेभ्योऽथ घोरेभ्यो अघोरघोरेतरेभ्यः।
सर्वतः शर्वः सर्वेभ्यो नमस्ते रुद्र रूपेभ्यः॥

Aghorebhyo thagorebhyo ghora ghoratarebhyaha
Sarvebhyassarva sarvebhyo namasthe astu Rudra rupebhyah.

One whose morph are darkest, the "blacker **than black**" material, virtually denser then densest form (This precisely means Aghora does not absorb, reflect or emits the light, Modern **Science** encompasses a large body of evidence collected by repeated observations and experiments about the behaviour of Dark Matter and that is Aghora).

सर्वतः शर्वः सर्वेभ्यो means he is everywhere modern science estimate it as 25% of cosmic structure (existence of dark matter as Aghora Shiva only inferable from the gravitational effect it seems to have on visible matter) Salutation to the auspicious **Shivaay** the semblance of every form, Salutation to that who is the aspect of all illusion but remain himself free from any illusion.

Siva's incarnation of **Aghoresha** looks *south* and of blue complexion representing destructive/ regenerative energy and Invocation to Siva states:

Aghorebhyo thagorebhyo ghora ghoratarebhyaha/ Sarvebhyassarva sarvebhyo namasthe astu Rudra rupebhyah.

Vande Dakshina -meeswarasya kutila bhrubhanga Roudram Mukham are Aghora states.

Then from him four individual of black featured was appeared. They all are the being with great with auspicious and supernatural powers called as Krishna, Krishnasikha, Krishnasya and Kanrishnakanthadhrika. For the sake of Brahma's creation, Lord Narayana appeared as the four Rudra deities as Krishna, Krishnasikha, Krishnasya and Kanrishnakanthadhrika the brought the miraculous yoga called Ghora, which seems terrible but pleasant by virtue.

Nandishwar Ji said:- The four son of Rudra Aghora performs distinct role in the universe. They all establish Aghore yoga.

Aghor is a simple and natural state of consciousness, in which there is no experience of fear, hatred, disgust or discrimination.

There is sect called Aghori (who practices the Aghore Yoga) who remains non-dualistic in all conditions.

Shital:- You have solved most mysterious theory of Dark Matter. The one of the most galling mysteries in physics is that of dark matter (Aghore) and dark energy (Kali), the names given to the unknown material and energy that observations suggest permeate the universe, but that we can't see. Scientists believe that together, these dark materials could account for up to 95 percent of the total mass in the universe.

I want to know more about the son's of Rudra Aghora, because as you described the names of Aghora's son's have the affinity and seem to be resemblance to Lord Shri Krishna, who is the incarnation of Lord Narayan Shri Hari Vishnu. He is the hero coryphaeus of Mahabharata age. Can you cast the ray of lights on this evidence?

Vijay:- Obviously Shri Hari Vishnu is the same Lord Narayana, the Krishna in Mahabharata. In fact he is not just the incarnation but his presence is in whole cosmos uniformly. Shri Krishna of Dwapar Yuga is the same Space, manifested to protect the principle of truth and morality. As we already discussed Lord Narayan is also one aspect of Rudra. Let me brief it in more pragmatic way.

The universe at this stage was with only very diffuse matter remaining with very low energy levels and **very large time scales.** miniature events happens during this period, and the universe is dominated by mysterious "dark matter". This is what the modern science says about, this Dark Age (**Kali** or Dark Era) remains from 300,000 to 150 million years

The same was narrated by Nandishwar Ji to Sanat Kumar Ji, the **very large time scales** means as a time **Brahama Ji** was lonely and he do not found anything other than himself and started to meditating on **Shiv.**

The scientific phenomena of Supreme Brahm are the epoch when the entire space was enwrapped by **Kali** (first Maha Vidya) to destroy the evil antimatters.

The origin of super massive black holes may remains an open question for modern scientist, with several competing hypothesis being put forth by them but it was well described in Shiv Purana by **Shri Nandishwar** Ji to **Sanat Kumar.**

After that, epoch was free from antimatter, Rudra Ahgora appeared in front of Brahma Ji, and helped him in the processes of creation.

Krishna, Krishnashikha, Krishnasya and Krishnakantdhak is the four descendants and successor of the Aghore Shiva.

They all are the epitome of Parbrahm and greatest master of Yoga. They are the regenerator of cosmic world.

Shital:- Obviously, when this phase of the universe has not found to grow then **Rudra Aghora** and his son's are the one who regenerated the universe after the distraction of Antimatter in association with Mother **Kali.**

Vijay:- Rudra Krishna emerged from Aghore and to be the one of the aspect of Black hole just after his appearance he asked to set his duties from his father **Aghore Shiv.**

Rudra Krishna asked:- O Lord of lords what is the purpose of my origin. Please indicate my duties.

Aghore Shiv Said:- Oh son, you are the cosmic person with great splendour and you have to help Shri Brahma Ji, in the processes of creation.

Though, because you have emerged from that's why your complexion is dark, but your nature is to enlighten the world with your power of great splendour.

Your magnificence glory will regenerate the entire universe while your another brother Krishnakanthak is having the efficacy of engulfing the material of the cosmos.

You are the elder among your brother's **Krishnashik, Krishnasya and Krishnakantdhak.** Thus you will assist and guide them to operate the universal laws.

Rudra Krishna asked:- Were I have to stay?

Brahma Ji Said:- Hey Krishna, Lord Shri Vishnu is having numerous mouth, hands, eyes and feet. He is a great cosmic man (Pursha) and you are non another then him, there is no difference in you and Lord Vishnu, the only thing I understand is that you are Rudra Krishna and you are the Narayan.

Sahasra sirṣa puruṣa | sahasra akṣaḥ sahasra pat |
Sa bhumiṁ visvato vrtva | atyatisthad dasangulam ||

So you please reside in the mouth of Virat Pursha Shri Vishnu while **Krishnashik, Krishnasya** stay in left and right nostril.

Your amity brother **Krishnakantdhak** will remain in the *glottis of Virrat Pursha Shri Vishnu.*

Shital:- O that's great, thanks for unlocking the truth, wisdom and the scientific knowledge of Vedic and Puranic texts and that is undeniable. How the secret are hidden in Vedas & Puranas, it's unbelievable up to now nobody can't even imagine that Devi Kali as a Dark-energy and Aghore Shiva as a dark Matter is exactly the essence of the power of Adipara-Shakti and Lord Shiv and the whole cosmos is their divine science to rule the universe.

It is Aghore Shiv and Maha Kali for which modern science is predicting as the theory of Dark Matter and Dark energy of Dark epoch of early universe.

And what I understand form your above narration is **Krishna, Krishnashik, Krishnasya and Krishnakantdhak** are respectively the holes associated with Dark Matter and Dark Energy. If I am right please elaborate the details of these.

Vijay:- Once the same has been asked by my friend Mr. Saktishwar Singh and Mr. E.V. Shiril in Coimbatore, I am going to tell you same what I told to him in the regards of Blackholes.

Shital:- As you told, Nandishwar Ji, said to Shri Sanatkumar Ji that this is the phase of Dark Age and the Lord started this journey from Brahm Swaroopa i.e.Nishkal to Sakal, from Sakal the *supreme* personality of *Godhead* proceeded to binary form of Shiv and Shiva i.e. Shiv and Adi Shakti by Anath Naad (Big Bang) then the processes of inflation begin as a vast Space of Maha Vishnu (Virat Pursha) after that the universe got the Time (The Brahama Ji).

Vijay:- We know now, that after the substantial manifestation of matter as a Rudra, therefore all the law's broke by the summon of Rudra's (different form of Matter) and Energy.

Scientifically the universe got four fundamental forces, the Primordial epoch of reionisation, first gaseous as hydrogen, then subatomic to atomic world and the Dark Age **(Kali).**

Now if we imagine that we are in the Dark Age, the age of Kali and we are watching the phase changes by the appearance of Aghore Shiva we will observe the appearance of Aghora and four Kumara's emerges from him. This is what we discussed above.

Now this is what the modern science and Shiv Puran refers on page no.420 about this play of Lord. We will go word by word meaning of the all the four individual.

Krishna – Black and auspicious.

Krishnashik – The flame or the pinnacle (Krishna means black + Nashika means hole which discharge white material). So literally Krishnasika means a White Hole.

Krishnasya – Krishna + Nasya ('Nasa' as a general term in Sanskrit means the "nose" and 'Nasya' as a treatment means to put something in the nose. So the term Nasya generally means a type of Ayurveda treatment that is done through the nose).

Krishnakanthadak – Krishna who hold the Mass inside glottis or neck.

A white whole **Krishnashik** is right nostril of **Virat Pursha Shri Vishnu**, Shiv Purana refers and the opposite nostril of Narayan is black hole as a **Krishnasya**. "A black hole **Krishnasya** is a place where the disolution or annihilation takes place and nothing can escape from him; a white hole is a place where **Krishnashik** stay and the universes emerge from him".

This is the cyclic phenomena of Lord, the material emerges from the **Krishnashik** the white holes and then the universe or material merges in Black hole **Krishnasya** of Virat Pursa after the completion of life cycle.

This is the yogic method or action called *Anulom Vilom or Pranayam* is an alternate breathing technique. For human being its merely breathing exercise to purify the mind and body by inhalation of oxygen and the expel of Carbon dioxide, but for **Shri Krishna** and, **Krishnashik and Krishnasya** it's the method of purification, dissolution and the creation of universe.

Anulom Vilom Pranayama or alternate nostril breathing exercise is one of the main practices of Pranayama. *Anulom Vilom* Pranayama is mentioned in the yogic texts *Hatha Yoga Pradeepika, Gheranda Samhita, Tirumandiram, Siva Samhita, Puranas* and in the *Upanishads.*

In the practice of pranayama, inhalation (called *Puraka*), retention (called *Kumbhaka*) and exhalation (called *Rechaka*) is used. Anulom Vilom pranayama can be practiced with or without *Kumbhaka* (holding of breath).

Now, this in case of cosmic creation, Lord Maha Vishnu or Virat Pursha is always in the state of yoga. **Krishnasya** as a Blackhole with his right nostril continuously inhale whole the celestial bodies and the cosmic material that has completed their life cycle, while **Krishnashik** who remain in left nostril inception of the new cosmic life of universe is a white-hole.

You might remember the episode of Virata Swaroopa of Mahabharata and TV serial Shri Krishna, or You may go through the text of Upnishads and Puranas which states that when Shri Krishna shown his Virat Swaroopa to Arjuna he saw the thousand of Stars, Planets and Galaxies are inhaled by his one side of nostril and expelled from other side of nostrils of Virat Vishnu Shree Krishna and the processes of inhalation and expiration of universe goes on. This process continues from his numerous nostrils.

While **Krishnakantdhak** stay in glottis (neck) has taken the charge of retention of Vishwa (Universe) i.e. **Krishnakantdhak** is the universe in the glottis of vast space or the Virat Pursha Shri Narayana. Hear Adi Shakti also act as a Singularity and everything whatever taken through the Black Hole by **Krishnasya** is consumed as a one in Singularity. The singularity is where mass of diffused matter is squeezed into a point of infinite density.

Rudra Krishna is the one who take care of transit, the transit of life as well as matter from one world to another universe, is the task performed by Shri Krishna as he operates the warm hole. Shri Krishna himself is the infinite universe.

Shital:- Oh, Amazing but one question counters on my brain as per the modern science there are numerous black holes and warm holes in the universe. What you have narrated about **Krishna, Krishnashik, Krishnasya and Krishnakantdhak** is that one or multiple?

Vijay:- Interesting, you might have remember our above discussion on **Krishna, Krishnashik, Krishnasya and Krishnakantdhak** was something around a year back, but today almost after a year a Astrophysical Journal provided by *Princeton University* is published on **March 13, 2019** at public domain www.physics.org.

This is the strong evidential support of common notion of science and Vedas. Let's take a glimpse of this journal.

Astronomers discover 83 supermassive
black holes in the early universe

Astronomers from Japan, Taiwan and Princeton University have discovered 83 quasars powered by supermassive black holes in the distant universe, from a time when the universe was less than 10 percent of its present age i.e. that were formed when the universe was only 5 percent of its current age.

"It is remarkable that such massive dense objects were able to form so soon after the Big Bang," said Michael Strauss, a professor of astrophysical sciences at Princeton University who is one of the co-authors of the study. "Understanding how black holes can form in the early universe, and just how common they are, is a challenge for our cosmological models."

This finding increases the number of black holes known at that epoch considerably, and reveals, for the first time, how common they are early in the universe's history. In addition, it provides new insight into the effect of black holes on the physical state of gas in the early universe in its first billion years. The research appears in a series of five papers published in The Astrophysical Journal and the Publications of the Astronomical Observatory of Japan.

Supermassive black holes, found at the centers of galaxies, can be millions or even billions of times more massive than the sun. While they are prevalent today, it is unclear when they first formed, and how many existed in the distant early universe. A supermassive black hole becomes visible when gas accretes onto it, causing it to shine as a "quasar." Previous studies have been sensitive only to the very rare, most luminous quasars, and thus the most massive black holes. The new discoveries probe the population of fainter quasars, powered by black holes with masses comparable to most black holes seen in the present-day universe.

The research team used data taken with a cutting-edge instrument, "Hyper Suprime-Cam" (HSC), mounted on the Subaru Telescope of the National Astronomical Observatory of Japan, which is located on the summit of Maunakea in Hawaii. HSC has a gigantic field-of-view—1.77 degrees across, or seven times the area of the full moon—mounted on one of the largest telescopes in the world. The HSC team is surveying the sky over the course of 300 nights of telescope time, spread over five years.

The team selected distant quasar candidates from the sensitive HSC survey data. They then carried out an intensive observational campaign to obtain spectra of those candidates, using three telescopes: the Subaru Telescope; the Gran Telescopio Canarias on the island of La Palma in the Canaries, Spain; and the Gemini South Telescope in Chile. The survey has revealed 83 previously unknown very distant quasars. Together with 17 quasars already known in the survey region, the researchers found that there is roughly one supermassive black hole per cubic giga-light-year—in other words, if you chunked the universe into imaginary cubes that are a billion light-years on a side, each would hold one supermassive black hole.

This journal state that in there observatory they found 83 black holes and many more are yet to be there to find out in future observation. In their research, the team also could not able to detect about the formation of 1st black hole.

Now the same with perfect details of event what cause 1st black hole **Krishnashik,** to come in extant was clearly mentioned in Shiva Purana. Shiv Purana not only certify the origin of 1st black hole, in fact it also endorsed about the origin of 'white' and 'warm' holes named as **Krishnasya and Krishnakantdhak** and there number are Shahastra (thousand) or in other words numerous black, white and warm holes are evolved as with the inflation of Space (Narayan)

Whole universe is truly the space, in the form of Maha-Vishnu who is having numerous mouths as referred in Purush Sukta of Rig Veda.

सहस्रशीर्षा पुरुषः सहस्राक्षः सहस्रपात् ।
स भूमिं विश्वतो वृत्वात्यतिष्ठद्दशाङुलम् ॥ १ ॥

Sahastra-Shiirssaa Purusha Sahasra-Akssah Sahasra-Paat |
Sa Bhuumim Vishvato Vrtva-Atya -Tisstthad-Dasha-Angulam ||1||

He Purusha (Universal Being) has Thousand Heads, Thousand Eyes and Thousand Feet (Thousand signifies innumerable which points to the omnipresence of the Universal Being).

He envelops the World from all sides (i.e. He pervades each part of the Creation), and extends beyond in the Ten Directions (represented by Ten Fingers).

Your answer is in the first sentence of Purusha Sukta which state that the Space or Universal Being lord "Narayan Shri Hari Vishnu" or 'Pratham Pursha has thousands of Heads (सहस्रशीर्षा).

Shital:- It means, the Space lord Vishnu has numerous heads thus there are numerous black holes, white holes and warm holes in the representation of **Krishna, Krishnashik, Krishnasya and Krishnakantdhak.**

Vijay:- Exactly, **Krishna, Krishnashik, Krishnasya and Krishnakantdhak** are the master of holes but during this early phase of the universe **Krishna, Krishnashik, Krishnasya and Krishnakantdhak** formed as a primordial black, white and warm holes, as the space expands and inflation goes on the generation of **Krishna, Krishnashik, Krishnasya and Krishnakantdhak become numerous.**

In the reference of latest research a **black hole** takes in matter, its gravity is so strong not even light can escape it. A **white hole** is the exact opposite, it ejects matter. However we haven't observed a **white** hole and a wormhole is like a tunnel from one side of the universe to another, or any other spot.

So Krishna pair up its black holes with *Jati-the gravity* and suck the matter of stars and even galaxies after the completion of their life span.

As Nandishwar narrated that, Rudra Aghora and his progeny leads the early universe from the phase of **Kali (Dark Age)** into the **epoch of reionization.**

Now we will discuss the Dark Age and the parley of Dark Matter with his progeny. We know that Goddess Kali has embraced every primordial thing of early universe within her. The only thing remains at Dark Age was the diffused Matter and these diffused matter was inhaled by Aghore and his son Krishnasya as a Black Hole.

Rudra Jati was assisting by his strong power of gravity to **Rudra Krishnasya** and in the demarche whole the diffused matter was inhaled with the strong force of gravity by Krishnasya.

The most mysterious thing for modern scientists is the Dark Matter but Dark Matter is not a mysterious matter in Vedic cult, it's the **Rudra Aghora** whose presence is approximately 27% of the mass of the entire universe.

There is a common belief in modern science that Dark Matter is produced by the earlier primordial black hole.

But the fact is that Black holes including white, warm and cool holes are begotten from the **Aghore Shiva** as a Dark Matter. The four individual are very particular in their duties in the epoch fusion or merge, retention and reionization of celestial objects.

There's the common notion that black holes **Krishnasya** suck everything in the vicinity by exerting a strong gravitational of **Jati** influence on the matter, energy, and space surrounding them.

But astronomers have found that the dark matter around black holes resists assimilation. There is no effect of Black hole over the Dark Matter because Dark Matter Aghora is the generator of Black hole.

The Aghore Shiva cannot be seen by ordinary eyes; even nobody in this universe can see him except Adi-Shakti the Dark energy and Lord Maha Vishnu because Aghore is the epitome of yoga and remains always in the state of deep meditation.

He can be only detected through its gravitational (**Jati**) influence on its surroundings.

We already discussed in detail about Aghora, now what he does at the earlier Dark Age is the main question. For which as I told you that, he is the one who generated four aspects of dark matter by the name of **Krishna, Krishnashik, Krishnasya and Krishnakantdhak.**

Now by taking a scientific predictions and prospective, the theory of Aghora can be like this.

There is excellent evidence that **Krishnasya** black holes really exist, and astrophysicists have a robust understanding of what it takes to make one. To imagine how a **Krishnashik** white hole might act, though we have to go out on a bit of an astronomical limb. Modern Scientists says one possibility involves a spinning black hole.

According to Einstein's general theory of relativity, the rotation smcars thc singularity **(Adi Shakti)** into a ring, making it possible in theory to travel through the swirling black hole without being crushed. General relativity's equations suggest that someone falling into such a black hole **(Krishnasya)** could fall through a tunnel in space-time called a Krishna as a wormhole and emerge from a white hole **(Krishnashika)** that spits its contents into a different region of space at different of time.

You know the evidence occurred during the Dark Age (Kali). At the end phase of Dark Age (Kali) by the processes of reionization, matter in the universe began to clump together and there is call of **Sunand, Nandan, Vishwananda** and **Upnandan Rudras by Lord Shri Krishna Narayan.**

They all together work along with the Aghoras and his son's thus the time came to the origin of first Star as a Tara Devi. So this is the processes of transformation of **Kali** (Dark Energy) to the first shining object **Tara Devi**. Tara Devi is considered as a 2nd wisdom among Dus Mahavidhya.

On the contrary, the 1st Maha Vidhya Kali, the Goddess appeared in same form with slightly different features according to the need of time for the welfare of world. She is before Anath Naadh (Big- Bang), She is the Dark epoch of universe before reionization. She has adapted the Bhadra Kali, the auspicious Swaroopa (aspect) during the war between **Veerbhadra** and **Daksha Prajapati**. She also adopted the same Kali a fierce fighter and terrible form to kill Raktabiza. We will discuss more about Maa Kali in further chain of tale.

Tara Devi emanation is the transition from darkness of Kali towards the light, we know that during inflation of Vishnu, dark energy **(Kali)** made the universe smooth out and accelerate.

Vijay:- The five aspects of Shiv and their progeny are the source, subject and matter of universe who operates the cosmos with energy Adi-Shakti in various forms.

Nandishwar Ji Said:- Hey Sanat Kumar Ji, this way for the weal of world I have narrated you the emanation of Sadyojata, Ishana, Vamdeva, Tatpursha and Aghora Shiva along with different forms of Adi Shakti.

In brief movement of galaxies, objects, stars, galaxies, nebulas, interstellar gas and everything seen and unseen structure of cosmos is operated by the distinct and prominent aspect of Rudra and they all engaged in the welfare of world.

5.10 The Emanation of Tara Devi (Appearance of the First Star of Universe)

Vijay:- As time went on, the Dark Matter as **Aghora** dominated the whole world during the Dark Age. Now during this phase the primordial hydrogen (Sadyojata and his sons) gas was cooled enough and when the **Krishnasya** the black hole suck the diffused matter, during this transition the matter collapsed by the gravitational instability of **Jati** from small initial density perturbation. As they assemble via hierarchical merging, primordial gas cools through ri-ionization lines by Sadyojata as hydrogen molecules and sinks to the center of the dark matter potential well.

Due to the influence of **Jati (Gravity)** the cooled gas becomes self gravitational as it result in the strong and rapid contraction between Sadyojata, Aghora and Dark-energy.

Though, the entire theory of Dus-Maha Vidhya's is mentioned in Vedas and Durga Saptsati. The ten wisdom are the basically the ten form of Adi-Shakti for the execution of cosmos.

The succession of Dus-Maha Vidhya's is an absolutely the theory of beginning, sustaining, growth, and distractive phases of the universe, the same was described by modern Science, can compare in the following manner.

- Kali - (Dark Age).

- Tara Devi - (1st Star of the universe).

- Tripura Sundari or Shodashi - (arrangement of "three worlds," "three planes of existence," "three realms" and "three regions).

- **Bhuvaneshwari** – (Large scale structure formation, the emergence and the evolution of formation of Galaxies).

- Bhairavi – (The Cosmic energy throughout the universe) the Fierce Goddess.

- Chhinnamasta – (split fraction of universe)The self-decapitated Goddess

- Dhumavati – The annihilation of the cosmos The Widow Goddess, or the Goddess of death.

- Bagalamukhi – The Goddess Who Paralyzes Enemies

- Matangi – the Prime Minister of Lalita (in Srikula systems); the "Tantric Saraswati"

- Kamala – (The goddess who sustain Universe)The Lotus Goddess; the "Tantric Lakshmi"

The primordial gas (Sadyojata) out of which the first stars were made, had 76% of its mass in hydrogen **(Sunand)** and 24% in helium **(Nandan)** and did not contain elements heavier than Lithium. This is because during Big-Bang nucleosynthesis, the cosmic expansion rate was too fast to allow the synthesis of heavier elements through nuclear fusion reactions.

As a result, cooling of the primordial gas and its fragmentation into the first stars was initially mediated by trace amounts of molecular.

It will be easier to understand as if we refer the modern science. Observations reveal that tiny clumps of matter formed in the baby universe; to WMAP (Wilkinson Microwave Anisotropy Probe (**WMAP**) is a NASA Explorer mission; these clumps are seen as tiny temperatures differences of less than one-millionth of a degree. Gravity then draged more matter from areas of lower density and the clumps grew. After about 200 million years of this clumping, there was adequate matter in one place that the temperature got high enough for nuclear fusion to begin - providing the engine for stars to glow.

Subsequently, the **Tara Devi** appearance became much more efficient through the cooling of **Sunand** (atomic hydrogen and Hydrogen bonding). The evolution of star **(Tara Devi)** formation began by the cooling effect of hydrogen, which was done by **Sadyojata** and his son **Sunand,** along with the influential behavior of **Jati,** the Rudra with the power of Gravity. Everything in the cosmos of that phase was genetically driven from Dark-Energy because when Rudra Try to approach Dark-Energy, the Dark Energy (Kali) repels with big thump to **Jati** which results in the transformation of **Dark Energy (Kali)** into the first shining Star called **Tara Devi.**

But before we get to the late stages, let us start with the beginning and examine the formation sites of the very first stars.

How did the first clouds of gas form and fragment into the first stars?

Generally, the collapsing region makes a central massive clump by the behavior of Blackhole **(Krishnasya)** with a typical mass of hundreds of solar masses.Which happens for a temperature of ~ 500K and the density ~ 104 cm−3 at which the gas lingers because its H2 cooling time is longer than its collapse time at that point. This is what the modern science says about the temperature.

Soon after its formation, the cluster of gas (Mass) becomes gravitationally unstable by the hit from Dark Energy because Dark Energy repels the gravity on **Kali** (i.e. on Dark energy there is no effect of gravity) and undergoes runaway collapse at a roughly constant temperature due to H2 cooling. The central cluster does not typically undergo further sub-fragmentation, and is expected to form a single star i.e. **Tara Devi.**

The origin of **Tara Devi** is the final era of the "cosmic dark ages." But Kali is the primordial goddess appears time to time for the destruction of evils.

Tara Devi- in the form of 1st star ruled the cosmos for thousands of years as per Vedic view. The Tara Devi is having a mass and radiation of billions of sun. She illuminated the earlier world soon after the end of Dark Age.

Though, this is also the epoch of the formation of heavy elements within the core of star (in the womb of Tara Devi).

The attribute of Tara Devi is described as blue as per Vedic and Puranic, which is supported by modern science which states that the mass and luminosity of a star also relate to its color. More massive stars are hotter and bluer.

Shital:- In appearance **Tara** Devi is blue while Goddess **Kali** appearance is black. But I want to know her relation with Shiva because her both forms (Kali and Tara) are also depicted as standing over Shiv?

Vijay:- This shows your zeal, interestingly you raise very valid question. As you know we have discussed earlier that Mass and Energy are like two sides of one coin. They are the Balance Symmetry of existence. Shiv and Shakti is aspect of equilibrium and the phenomena of libration.

The almighty appears in many wonderful form like Mass, Energy, Space and Time. He appeared in the form of matter as a Rudra, God himself transform to reciprocate the proper functioning of cosmic world in the form of atomic, subatomic particles, light and heavy elements. Gaseous and dust are the form of Sunand and Nandan.

He transforms as a black, white and worm holes by the aspect of Rudras, on the contrary he elicited forth the *Jati, Mundi, Ardhmundi and Sikhindi* as the four fundamental forces of the universe.

It should be noted that the Modern and Vedic Science view in this regard is also the same, first quasars (region surrounding a supermassive black hole or Krishanasya) form from gravitational collapse, and the intense radiation they emit re-ionizes the surrounding universe, the second of two major phase changes of hydrogen gas in the universe (the first being the Recombination period) this all has been generated at the core of **Tara Devi**. From this point on, most of the universe goes from being neutral back to being composed of ionized plasma.

The scientific facet of this is that the Devi Kali and Tara Devi both are very massive and the Vedic cult Say's its only Lord Shiva, the one who can control the aggression of Shakti, only he can pacify the physical and psychological force of fiery attribute of **Kali** and **Tara.**

When the aggressive behaviour of Goddess is uncontrolled, the mass starts to please the energy. This is what we see in the pictures or photo images in which Lord Shiva laying supine under the feet of Devi Kali.

It is commonly accepted estimate was that, these early stars were hundreds of solar masses, the researchers said. After ruling for big span **Tara Devi** has generated many elements at the core.

She decided to transform herself to give a new shape to the universe. Slowly she transformed herself by a big explosion, in turn given chemical creations into outer space, making the way for subsequent generations of stars, solar systems and galaxies.

With a greater understanding of Vedas our Rishies have simplified the Vedas into Puranas and then again the great work done by earlier generation of saints like Rishi Markandaye which was followed by the creation of Upnishades and all these secrets of chronology of universe is decoded in these ancient literatures.

Goddess **Tera Devi** is also called the **Neela Saraswati** because of her complexion and it's her energy pouring the knowledge to the world, the model, the life of a primordial supermassive star when fitted in modern scientific view.

As the first star collapses, it begins to rapidly synthesize heavy elements like oxygen, neon, magnesium and silicon starting with helium in its core.

This process releases more energy **(Tripura Sundari or Shodashi)** than the binding energy of the star, halting the collapse and causing a massive explosion: a supernova the birth of several new stars. This super energy is the Goddess called **Tripura Sundari.**

5.11 The Period of Large-Scale Morphostructure by Tripura Sundari or Shodashi - Galaxy Epoch

Vijay:- This is the epoch of the formation of all heavy elements, the formation of planets and galaxies. The Vedic and Puranic description is as:

Shital, listen we will start the discussion by chanting and recalling of Lalita Tripura Sundari Mantra.

"Om Aim Hreem Shreem Shree Lalita Tripurasundari Padukam Poojayami Namah"

The above is the simple mantra which is as we know **Aim** means Aimkari (Singularity), **Hreem** is a mantra of purification and transformation. Hrim brings transformation from one state to another by mean of energies. It is also the mantras of Devi Maha Maya,

Shreem is the seed mantras for prosperity, wealth, purity, and generosity or Devi Laksmi used in Shree Yantra Pooja i.e. worshipping the Goddess in the form of the Shree Yantra. Shree Yantra is the cosmic web in the form of different energy levels.

This is also regarded as the worshiping of Lalita Tripura Sundari and her different form of energies level through the universe. Tripura Sundari literally means the "Beauty of the Three Worlds"

We now know that Tara Devi, the first Star transform her energy as a Tripura Sundri and her energy remain intact in all the stars of universe.

Tripura literally means as:

Bhur – Prathvi

Bhurvab – Antriksha or the Sky

Swah or Swarga – The heaven or the heavenly bodies resides.

After the transformation and the collapse of first Star, the birth of many celestial bodies began which results the formation of 2nd generation Stars and Galaxies formation.

Therefore *Devi Tripura Sundary* means who made these beautiful triple worlds.

Devi *Lalita Tripura Sundary* (the one who indulges in the play to made the triple world beautiful), also referred to as Shodashi (the Vermilion-hued One)

Physical attributes Lalita Tripura Sundari is depicted as the energy level in sixteen aspects and here the cosmos are now having sixteen set of planets (another meaning for Shodashi), thus embodying the sixteen types of desire.

She is described as having a dusky complexion because after the transformation of Tara Devi (the very 1st star of cosmos) there was also the cluster of dust particles which make the early universe dusky.

An esoteric interpretation is that her body is said to be made up of the collective Shaktis and she is the energies of Brahma, Vishnu, and Rudra i.e. she is Brahmani, Vaishnavi and Rudrani respectively.

Shital:- That's great, to know the Swaroopa of **Devi Lalita Tripura Sundary** but I want to know her relativity with modern hypothesis.

Vijay:- See whole processes is the play of Lalita Tripura Sundari and Tara Devi to create the universe full of Stars and Galaxies.

The modern science called it, the "theory of nebular hypothesis", stars form in massive and dense clouds of molecular hydrogen—giant molecular clouds (GMC). These clouds are gravitationally unstable, and matter coalesces within them to smaller denser clumps, which then rotate, collapse, and form stars. Star formation is a complex process, which always produces a gaseous protoplanetary disk propelled around the young star.

Thus the formation of planetary systems is the thought processes of **Tripura Sundary** (the Energy which binds the Mass) result of star formation. A Sun-like star usually takes approximately 1 million years to form, with the protoplanetary disk evolving into a planetary system over the next 10–100 million years.

The similarity which we found is that, the **Sunand** (Molecular-Hydrogen), which are unstable by the influence and the behavior of **Rudra Jati** (gravitationally unstable) and the Matter left by **Tara Devi** coalesces within them to smaller denser clumps which then rotate, collapse, and form stars.

Lalita holds five flower arrows, noose, goad and bow. While the noose represents attachment; the goad symbolizes repulsion; the sugarcane bow, the mind; and the flowery arrows, the five sense objects. Devi Tripurasundari combines Kali's determination and Durga's charm, grace, verve and complexion. She has a third eye on her forehead.

She carries in her hands various attributes associated with Shiva. An aura of royalty characterizes her overall bearing and ambiance.

The Lalita Sahasranama describes the deity as extremely merciful, leading the devotee to liberation. She is the one who operate the three worlds of Stars, Planets and a Galaxies with various celestial bodies.

In the series of cosmic structure formation, this is also the one epoch of galaxy formation, the events follow one billion years before the oldest Galaxy HCM-6A known to the modern cosmology.

Formation of hyper-luminous quasar SDSS J0100+2802, which anchorage a black hole with the mass of 12 billion solar masses one of the most massive black hole **Krishnasya** discovered so early in the universe.

HE1327-2326, population II star, speculated to have formed from remnants of earlier Population III stars i.e. of **Tara Devi.**

Re-ionization completed—the Universe becomes transparent again by the grace of Goddess **Lalita Tripura Sundary.** Galaxy evolution and the formation continues because the Universe is still small in size, galaxy interactions become common place with larger and larger galaxies forming out of the galaxy merger process. *Devi Tripura Sundary* begun clustering and creating the largest structures in the Universe so far - the first galaxy clusters and galaxy superclusters appeared.

Goddess **Lalita Tripura Sundary** created the first galaxy of cosmos. We can also call even the first galaxy by the name of Tripura Sundary Galaxy. The processes of expansion of the universe were continued by Maha Vidhya Lalita Tripura Sundary.

Shiva Said to Narayan:- Hey Narayan, now the world is because of your grace is in Retention mode and I think Shri Brahama Ji's is free from all his worries,

Shiv further Said:- Hey Devi Lalite, now it's the period to take the universe to expand in to multi-verse.

Devi Adi Shakti Said:- Oh lord, your order will be mandate and I will transform some of my energy to create numerous galaxies.

Narayan said:- Hey Devi, by your grace only the multi-verse will be created, and by doing so you will be known by the name of **Bhuvaneshwari.** She is the creator of the entire universe.

Shri Narayan Vishnu recited the following mantra in the honor of Devi:

> **"Bhuvanesheem Mahamayaam**
> **Sooryamandalaroopineem"**

5.12 Eternal Inflation or the Theory of Multiverse as Bhuvaneshwari

Now at this stage of cosmic enhancement, there are multiple universes (Bhuvan) resting in the **Garbhodakshayi Vishnu** and Devi **Bhuvaneshwari** is the Mistress of the entire creation.

So **Bhuvaneshwari** precisely means Bhuvan + Ishiwari, i.e the one who is the Ishiwari (Goddess) of Bhuwans or the Goddess of infinite universes.

The multiverse is the designed phenomena by Shiv-Shakti, Maha Vishnu and the Viranchi Shri Brahma Ji.

Were the Rudra's are the prime part of planed universe in its multiverse form. So the fact is that we live in a multiverse containing an immense number of universes.

Rishi Markandya Says, The expansion of Space – Shree Vishnu Narayana required the energy and matter to grow in steady fast region and also to sustain the multiverse. This was done by Devi Maha - Maya in the form of **Bhuvaneshwari.**

That's why modern science says, that the Dark Energy (Goddess Kali) made universe smoother and she is in the cabinet of Dus Maha Vidhya's. While **Bhuvaneshwari** is fundamentally the Adi Mahavidya similar to (Kali, Tara and Tripura Sundari based on their role in universe).

Shital:- So Devi **Bhuvaneshwari** refers the cosmos with many galaxies and even as of now what I understood is that she rules over infinite universes.

Vijay:- Obviously, This is truly the playful scientific cosmological tale of Devi Adi-Shakti and Lord Shiva with Narayana.

Tara Devi is conspectus and proceeded with the plan on big scale to transform into **Bhuvaneshwari.** Stars are organized into galaxies, which in turn form galaxies groups, galaxy clusters, superclusters, sheets, walls and filaments, which are managed by **Bhuvaneshwari.** She creates a vast cosmic structure. She embodies the physical cosmos and the one who gives shape to the creation of the world.

In modern science it is sometimes called the "cosmic web". It was commonly assumed that galaxy clusters were the largest structures in existence and that they were distributed more or less uniformly throughout the Universe in every direction.

Same what Vedas claim that, **Bhuvaneshwari** the sovereign Queen or *Rajrajeshwari* is the fourth Mahavidya; As Mahavidya, her realm *Bhuvana* includes not merely one universe, but also the multiverse that means the entire creation. She is more related to the dynamics energy of the visible world that we experience and also that is beyond our perceptibility and knowledge.

Indeed the world is her extension. It is said; the universe is her body or **soma** (*Visvaswarupa*) and the entire world is the soul *blossoming* of her nature (*Sarvarupa*). She embodies all the liveliness and attributes of living nature. She represents the forces of the material world. The whole of existence is the field of her joyful play (Lalita); she is *Sarveshi* the ruler of all. She is also *Mahamaya* the great enchantress.

Devi Bhuvaneshwari expands universes into 14 Bhuvan's (Universes) thus she is the primal energy who created fourteen Galaxies (भुवन), each Bhuvan consist of distant planetary system and therefore each distinct Galaxy possess many solar systems.

Solar System is a group of members of some Planets and certain Lokas. In the vast Space (Maha Vishnu) there are innumerable Vaikuntha, and the galaxy of the supreme Lord from where the expansions of the Supreme Personality of Godhead come into this universe.

The supreme primordial energy in the form of *Bhuvaneshwari* is the goddess, who feeds all the universes through the nectar of her power.

The Narayana holds and take care of all the planets (Loka) and Galaxies (Akaas Ganga or *Milky Way, Andromeda galaxy, whirlpool galaxies* etc.) in the same way as we take care of our children.

Shital:- How these Lokas and planets are arranged in the early phase of universe. As you told, the **Bhuvaneshwari** Maha-Vidhya is the form of Adi-Shakti, who wears the universe, came into *existence* in the early part of cosmic structure formation. What exactly is the difference between Lokas and Graha (Planet)?

Vijay:- I will explore the cosmological event taken place in the pre historic time. This is the Vedic & Puranic science, as you know the continuation of universe expansion still goes on. You should also note that in this vast multiverse Devi **Bhuvaneshwari** Maha-Vidhya has arranged each universe with the sets of beautiful Galaxies. The account of Galaxies in each universe varies and differs in size, amount, degree or nature of existence overthere. At the initial phase of cosmos most of the universes have set of fourteen galaxies as space expands these galaxies multiplied in manifolds.

Arrangements of Galaxies and Planets by Brahma Ji

Vijay:- Now, I anamnesis (remember) the story of Vishnu Purana in which a brief description of planets is given. I will narrate the same ancient science. The arrangement of Lokas up to the narration of this story was of fourteen Loka's.

As you asked Loka's are not just the celestial bodies but it states the location of particular living planets.

From the time of galaxies formation to present expansion of the universe was so vast that is from fourteen galaxies to enormous galaxies come in existence till the present.

Once the sage *Maitreya* came to the sage Parashara and wanted to know about the planets or **lokas** of the universe. Sage Parashara replied.

Parashara Said:- O Maitreya Ji, The Lokas are arranged in vast Shri Vishnu (Space). There are multitudinous universes and every distinct universe with multiple galaxies and in each there are some 14 Loka's.

The 14 Lokas –

The concept of the 14 Lokas of is the state that they are divided into 7 upper worlds or Vyarthis and the 7 lower ones, known as the Patalas.

The 7 Vyarthis –

1. **Satya-loka:** Brahma's loka. Satya-loka planetary system rich in Satava Guna is not eternal. Abode of Truth or of Brahma, where Atman (Shri Narayan) is released from the necessity of rebirth through Brahamaji. The location of this Loka is in the Galaxy called *"Dhavala"* from where the master of time Shri Brahama indulge in the processes of creation and this is called to be the first Galaxy of the universe with enourmous material. The Adi Shakti in the form of "Maha Saraswati" is the companion of Shri Brahama Ji, as of now many such Satya Loka was created with the expansion of the universe in extragalactic region.

    Brahma's, "Cosmic Clock" has started with the original Big Bang. Quantization of time (Brahma) is a key to explain the working of the universe as a computer simulation at different dimensions. Each quantum cosmic location and cosmic second form a unique digital "Spacetime" code which forms the index for the "Cosmic Record" associated with every interaction of a subatomic Particle.

    Hence the formation of Dhavala Galaxy begins just after the supernova of 1st Star i.e. Tara Devi of the universe, which leads to the formation of numerous 1st generation stars so-called population III **stars** and each star again posses the same energy of Adi Maha Vidhya Tara Devi. Dhavala galaxy is the oldest known galaxy of Vedic Cosmic science.

2. **Tapa-Loka:** This Loka is the arrangement of Stars who achieved this state by thire great penance. Like "Dhruba Loka" (Polar Star and its associated group of stars) has its distinct place in the space. Same way Tapa loka was created to achieve this zenith by the result of penance, truthfulness and devotion in God because god is the ultimate reality.

Vijay:- Tapa Loka is known as a Polaris Galaxy is a galaxy which is located at the center of our universe, which is the edge of the known universe as stated both by Vedic as well as modern astronomy.

It has a very close dwarf binary, and a larger star, Polaris B, which **orbits** 2,400 AU away. The main star, Polaris A, is a giant with 4.5 times the mass of the Sun and a diameter of 45 million Kilometers. It is a classic Cepheid variable, the closest to us in the whole **Milky Way.**

3. **Jana-Loka:** Abode of the sons of God Brahma. It is also believed to be the abode of the Hindu god Brahma's sons Sanaka, Sananda, Sanat and Kumara in some traditions. It's the plane of liberated mortals.

4. **Mahar-Loka:** These are also the habitable location, where abode of great sages and enlightened beings like Markendeya and other Rishies.

5. **Svar-Loka:** Region between the sun and polar star, the heaven of the god Indra. Indra, devatas, Rishies, Gandharvas and Apsaras live here: a heavenly paradise of pleasure, where all the 330 million Hindu gods (Deva) reside along with the king of gods, Indra.

6. **Bhuvar-Loka** (aka Pitri Loka): Sun, planets, stars. This is the space between the earth and the Sun. It is a real region, the atmosphere, the life-force.

7. **Bhur-Loka:** The Vishnu Purana says that the earth is merely one of the thousands of billions of inhabited worlds like itself to be found in the universe.

The 7 Patalas –

Vijay:-Below description is also available for your quick **reference from** Shrimad Bhagwatam **2:5:40-41.**

Lord Shri Narayan Said:-

dyu-pataya eva te na yayur antam anantataya

tvam api yad-antaranda-nicaya nanu savaranah

kha iva rajamsi vanti vayasa saha yac chrutayas

tvayi hi phalanty atan-nirasanena bhavan-nidhanah

SYNONYMS

tat—in His; *kaṭyām*—waist; *ca*—also; *atalam*—the first planetary system below the earth; *kḷptam*—situated; *ūrubhyam*—on the thighs; *vitalam*—the second planetary system below; *vibhoḥ*—of the Lord; *jānubhyām*—on the ankles; *sutalam*—the third planetary system below; *śuddham*—purified; *jaṅghābhyām*—on the joints; *tu*—but; *talātalam*—the fourth planetary system below; *mahātalam*—the fifth planetary system below; *tu*—but; *gulphābhyām*—situated on the calves; *prapadābhyām*—on the upper or front portion of the feet; *rasātalam*—the sixth planetary system below; *patalam*—the seventh planetary system below; *pāda-talataḥ*—on the bottom or soles of the feet; *iti*—thus; *loka-mayaḥ*—full of planetary systems; *puman*—the Lord.

My dear son Nārada, know from me that there are seven lower planetary systems out of the total fourteen. The first planetary system, known as Atala, is situated on the waist; the second, Vitala, is situated on the thighs; the third, Sutala, on the knees; the fourth, Talātala, on the shanks; the fifth, Mahātala, on the ankles; the sixth, Rasātala, on the upper portion of the feet; and the seventh, Pātāla, on the soles of the feet. Thus the virāṭ form of the Lord is full of all planetary systems (Shrimad Bhagwatam **2:5:40-41**)

1. **Atala-Loka**: The inhabitants of Atala Loka possess mystical powers.

2. **Vitala-Loka**: Vitala is ruled by the Rudra Hara-Bhava – a form of Shiva, who dwells with attendant *Ganas* including ghosts and goblins as the master of gold mines. The residents of this realm are adorned with gold from this region.

3. **Sutala-Loka**: Sutala is the kingdom of the pious demon king Bali.

4. **Talatala-Loka**: Talatala is the realm of the demon-architect Maya, who is well-versed in sorcery. Shiva, as Tripurantaka, destroyed the three planetary systems of Maya but was later pleased with Maya and gave him this realm and promised to protect him.

5. **Mahatala-Loka**: Mahatala is the abode of many-hooded Nagas (serpents) – the sons of Kadru, headed by the *Krodhavasha* (Irascible) band of Kuhaka, Taksshaka, Kaliya and Sushena. They live here with their families in peace but always fear Garuda, the eagle-man.

6. **Rasatala-Loka**: Rasatala is the home of the demons – Danavas and Daityas, who may be mighty but cruel. They are the eternal foes of Devas (the gods). They live in holes like serpents.

7. **atala-Loka**: The lowest realm is called Patala or Nagaloka, the region of the Nagas, ruled by Vasuki. Here live several Nagas with many hoods. Each of their hoods is decorated by a jewel, whose light illuminates this realm.

Vijay:- Above mentioned narration is the highlights how the Lokas are described by ancient seers in Vedas and Puranas. I narrated you just to understand the basics. I will reveal the exact scientific notion in subsequent period.

The references given by **Alan Watts**, a professor, graduate school dean and research fellow of Harvard University, drew heavily on the insights of Vedanta. Watts became well known in the 1960s as a pioneer in bringing Eastern philosophy to the West. He wrote:

"To the philosophers of India, however, Relativity is no new discovery, just as the concept of light years is no matter for astonishment to people used to thinking of time in millions of Kalpa's, (A Kalpa is about 4,320,000 years). The fact that the wise men of India have not been concerned with

technological applications of this knowledge arises from the circumstance that technology is but one of innumerable ways of applying it."

5.13 The Dwelling Bhairavi and Decay of Antimatter

Vijay:- By the summon of Rudras, the universe was full of Matter and Antimatter, we have had a detail discussion on matter in the form of Rudras in tangible and materialistic form, now we will see how the war won by matter with antimatter. The Goddess Bhairavi has the key role in the war between the matter and antimatter.

Shital:- Some unsolved mystery of modern science are always puzzling us. Modern science say's the big bang produce both matter and antimatter equally. Why is the universe made of matter and not antimatter?

Is there any Vedic proof or evidence which can veer the line of modern science? Is there any Puranic cult which can stand in support of this theory of matter and antimatter relation?

Vijay:- Partially, the literature of modern science is proofing in their own way and they are on the same track, but they have to understand the world of reality and optics. Modern theories are based on their perception and observation, while in Vedic and Puranic cults, theories are based on the evidence whatever the ancient seer saw in their time.

The goodness of Vedic philosophy is that it can be a melody or Vedic lore. It was taken by many, just as a song to praise Deities but no, it's a realistic science which can be recited as a song.

We can find much more detail in Vedas and Puranas but first need to understand the Sanskrit language.

There was nothing like imagination in the deliberation of dialogues in our discussion till now in the view of the timeline of the universe; Almighty pleased me to grant one to one signals in connection of Vedas science. This is not something called science of fiction but it's the real science which was narrated by Lord Shri Viranchi Brahma

Ji to Narad Ji, same theories are described by Nandishwar Ji to Shri Sanat Kumar Ji.

Moreover, many secrets are decoded by Lord Shiva himself and some wonders are disclosed by Shri Narayan Vishnu Ji.

Shital, now listen in the continuation of our discussion of Dus Maha Vidhya and the journey of the universe from finite to infinite, from Nishkal to Sakal, from Anhad Naad (Big-Bang) to the epoch of expansion of the universe was brief course of our discussion.

Devi Mahatmya gives us more details of Adi Shakti in her attributes of *Dus Maha-Vidhyas.* This follows completely the theory of transformation of energy. The frequency of *manifestation* of Adi Shakti and the transformation and manifestation of energy depends on the circumstances of the universe.

She created herself to abandon the miseries of cosmos; **Adi Parashakti** is the primordial cosmic energy and represents the dynamic forces. She is always active in her state of functional and structural form to run proper cosmic order of the system.

Now we will see the Vedic theory of further Maha-Vidhyas among ten Maha-Vidhyas and her role with matter and antimatter.

There are some more questions which modern scientists are unable to answer say.

- Why there was a war between Matter and Antimatter?

- How annihilation of antimatter was approached by Matter?

- Is further antimatter was produced after its annihilation?

The Vedic prospective is having all details of these anecdotes. I will disclose the ancient science of Vedas, which are remains as the enigmas for modern Physicists and scientists. My desire to explain **the** clockwork of God's creation is in the same with the way of modern physics.

Coming to the **Bhairavi, the Goddess of decay,** she is the cosmic energy, the fierce goddess but literally graceful for those on the righteous path. For those who has achieved the state of **Bhairavi,** is beyond the fear of death, sufferings, misery and pain and therefore awesome and librated.

Bhairavi is the energy present everywhere in the cosmos. She is the mother goddess who protects devotees and the righteous person from all the intensities, bringing joy in life.

She is the consort of Shiva in the form of Bhairav, Bhairav and Bhairavi both together are the legend of protector.

Her role in the universe is to decay the by product formed by the activation of Tri-Guna Satva, Rajsa and Tamsa, Guna remains in the state of equilibrium in Mulaprakriti, which has activity (Kartrittva), but no consciousness (Chaitanya).

The Antimatter behaves just opposite to that of Rudras and is the destructive morph for the creation.

Since Rudras are the the combined form of Triguna (Satva, Rajas and Tamas) and execute their duties to balance the universal matrix along with various aspects of Adi-Shakti.

Modern Science Says, that Matter and Antimatter are exposed to the universe is in equal amount with Big-Bang explosion, its right but not absolute truth.

Absolute realism is what the Vedas claim that with Anhad-Naad the only transcend balance image of the universe appeared and that is metaphysician for all the religion.

The cosmos immediately after Anhad Naad (Big Bang) was extremely pure dominant and transcend. The parting (Birth) of **Jati** from **Ishan** is to bliss the weal for the cosmos. But there was also simultaneous displacement of impurity began to start just after the phase of Grand Unification Epoch, from 10^{-43} seconds to 10^{-36} seconds, which was very less when compared to the noble Matter.

That all was the beginning of antimatter because of the separation of gravitational force that was narrated by Nandishwar Ji to Shri Sanat Kumar Ji as:

Nandishwara Said:- He Brahama putra, The force of **Jati** (Gravity) separated from the Maha Rudra **Ishan** (gross fundamental forces). **Jati** is the representative of attraction or *Akarshan* in respect of *Karna* (ear), *Vaani* (speech) and *Akaas* (space) closely connected with the frequency of sound. Thus with the origin of vast force, the act of attraction was also evolved and the birth of Anti-Particle started.

Which remain unified, and the earliest elementary particles and antiparticles begin to be created.

Gravity (jati) is somehow the attraction and when one particle attracts other particles, results in the collapse and collapse generates the sound and then sound traveled in the form of waves into Akaas (Space) which can be heard. The hearing or audibility is the attribute of ear.

So this was the seeds of Anti-Matter was poured in the cosmos at this epoch. This is the fiction science of the origin of antimatter.

But antimatter during that era was in dormant phase and was just nothing in front of the wellness of **Jati** and his brother's.

Nandishwar Ji said:- Hey **Sanat Kumar Ji,** during the epoch of Vamdeva Rudra and his son's **Viraj, Viwah,** *Vishoka and Vishwabhavan* are the master of compounds, they themselves are the first compound created by Rudra Vamdeva. Thus Vamdeva is also the Guardian Deity of ego (अहंकार).

It doesn't mean that they are the deities with egoism, in fact the remnant behind was the free radicals in the cosmos which was attracted by the power of attraction of the remnant of **Jati.**

The same was narrated by Lord Shiva to Devi Parvati that she is the mistress of protection and decay too. She is the combined form of trinity goddess (Devi Parvati, Devi Lakshmi and Devi Saraswati)

Shiva Said:- Hey Devi, you are the ह्रां ह्रीं ह्रूं भैरवी भद्रे भवान्यै ते नमो नमः.

In all those aspects, she decays the antimatter and protects the nature to sustain the cosmic order.

Vijay:- These sequel behave just opposite to that of Rudras (Matter) and that antimatter is the opposite of normal matter. More specifically, the subatomic particles of antimatter have properties opposite those of normal matter. The electrical charge of those particles is reversed. So this story is just opposite of matter.

The Vedic language is Sanskrit which is more progressive than any other language, according to which every objects are analogue of varying combination of Tri-Guna (Satva, Rajas and Tamas).

Satva as the Buddhi (intellect) particles for individuals and in universal form it is called **Mahat**, Rajas as an energy particle and Tamas is analogue of Mass particles.

That's why in our early discussion I have coded that Tri-Guna has the four phases Nishkal, Sakal, Sukasma and Purna (unmanifested, manifest, subtle and gross).

The first state is the state of Nishkal in which Satva, Rajas and Tamas are in equilibrium. Modern science states, the stage before big bang and the modern science does not have more idea about that.

From unmanifest state the Guna's are stirred into action by the effect of Rajas energy.

Nandishwar Said:- Hey Sanat Kumar Ji, In the case of Lord Shiv or Maha Rudra, he is the ascetic, mostly remains in the state of **Samadhi** (ecstasy) the deep state of yogic meditation.

He is the Lord of bringing balance to the universe, in the universe that is there is no effect of Tri-Guna on Shiva while Lord Vishnu is the master of **Tri-Guni Maya.** Entire world is enticed in the illusion of Maya Devi.

Shital:- Vedic science have complete clarity and it's more advance then western science. So Rudras are as:

For matter (Rudras), In the heart of an atom (परमाणु,कण), have the nucleus, posses protons the Satva Guna (which have a positive electrical charge) and neutrons (which have a neutral charge). Electrons, which generally have a negative charge, occupy orbits around the nucleus.

Vijay:- The matter has already been discussed by us in details by the realm of Tri-Guna (Satva, Rajas and Tamas). So in case of Antimatter the Tri-Guna behave Reverse, against, inverted and opposite to that of Satva, Rajas and Tamas. So this is formulated as a theory of 'enantiodromia'.

The Vedanta say's, those are antimatter at the epoch of re-ionization and later on manifested as Danava from Diti the mother of Daityas (demons).

"Antimatter is not antigravity," NASA added "Although it has not been experimentally confirmed, existing theory predicts that antimatter behaves the same way to gravity as does by normal matter."

The Vedanta say's,

As we know Satva is the character of Narayana, Rajas belongs to Brahama Ji and Rudra Shiv Shankar is the only one having retentive power of Tamas. While the absolute truth is that Shiva is the one who is the generator of all three Guna and he himself remains unaffected by the quirk and behaviour of all three.

Now we will come to the course of discussion that how the Antimatter broke a war with Matter and what does the Bhairavi do with that.

Nandishwar Said:- Hey Sanatkumar Ji, Goddess Bhairavi is the consort of Bhairav. The **Bhairava** (Sanskrit word, "Terrible, Frightful") is a fierce manifestation of Shiva associated with annihilation of evil Spirit, particles and Antimatter.

Lord Shiva Said:- Hey Brahama Ji, I and Shri Narayan will appear and incarnate in various forms to safeguard your creation, your creation will be protected by my Kaal Bhairav.

This Devi Bhairvi will destroy and decay all, whatever will be the negative spirits and evil **Danavas**.

Both Kaal Bhairav and Devi Bhairavi will roam throughout the universe to safeguard the world from evil spirits.

Shiva further said, Oh creator Prajapati Brahama Ji, Kaal Bhairav will appear time to time at different phase of the cosmos to annihilate the antimatter.

Vijay:- The remnant of these antimatter during each epoch from the time of re-ionization (During transformation of energy and Matter some remnant adopted the property of antimatter) decayed by the supernatural energy Bhairavi.

Bhairavi and **Kaal Bhairav** is the wandering form of Lord Shiva and Shakti and they guard the cardinal virtue of world throughout the existence.

Further generation of Kaal Bhairav was proceeded by eight Bhairava's are called Ashtanga or the Ashta Bhairava's (Ashta means eight). The **Ashta Bhairava's** control the 8 directions of each universe. Further the 64 (Sixty four) dimentions of the cosmos are safeguarded by 64 Bhairav's and are grouped under 8 (Eight Bhairava). Thus they are also called *KSHETRAPAL's one who protects the area.*

In further canto's it will be interesting to discuss the birth of antimatter and its annihilation by matter or else destroyed by energy or some time by lord Vishnu in order to protect the cosmos.

When the materialistic symptoms will be acute, intolerable and the fall of righteousness, the **Adi Shakti** or Lord **Shiva** or the **Narayana** appears in their transcendental power of various forms to repair the universal loss.

Shital:- In the course of our discussion you have disclosed so many hidden aspect of Vedas and Puranas, you have extremely enlighten my senses. This anecdotal description will definitely elaborate the senses of modern scientific scholars and modern generation.

Vijay:- In the continuation of our discussion about the origin of Brahmanda (Universe), we adopted the process of analyzing the recorded phenomena. I understand everything in this world was brought up by supreme almighty Brahm.

It's his allure to whom he want to realize his divine aspect otherwise every being is in ramble hallucination.

Now in the relation of remaining Maha Vidhyas, I will put the ray of lights in brief because these are the Maha-Vidyas which is the functionary aspect of habitat and re-constructive aspect of the universe.

The 1st five Maha Vidyas out of ten Maha Vidyas are the aspect of structural formation of the universe. In fact last five are the aspect of habitable universe.

The universe up to this epoch was fully matured, it was torridly burning that generated the heavier elements and populated the universe with full of Stars. The entire generation and the population of stars are by the energy of Tara Devi and fusion by **Tara Devi** (Star) in her core meld light atoms into heavier ones, all the carbon, oxygen, iron, and everything else needed to make dusty clouds, planets and the life. These light and heavy elements are conceived as an aspect of Tara Devi with the help of **Rudras** and placed in Space the Narayana called **Shri Garbhodakshayi Vishnu**.

These heavier elements are scattered around the early cosmos, when a star ends its life and explodes. So all the gas that exists in the universe now has a smattering of heavier elements, which allow it to cool more easily. As a result, stars tend to be smaller, burn less brightly, and live longer than their ancient forebears.

5.14 The Maha-Vidhya, Chhinnamasta Devi Habitable Epoch of the Universe

Shital:- The chronology of the **universe** describes the **history** and future of the **universe** according to Big Bang cosmology. But you allure the entire history from annihilation of the cosmos to the recreation of universe. Now I understood how the world evolved by the various aspect of Shiv and Shakti, How Narrayana become the ultimate truth and bliss of Anandam. How Nishkal turned to Sakal Shiva from the state of Singularity and Shri Vishnu becomes the vast perpetual universe. Is it there any Hymnal Stanza by which we can praise the creation of lord **Shiv** and **Shakti?**

Vijay:- The modern cosmology is divided into four parts of chronology of universe.

That is the Very early Universe, the early universe, the period of large-scale structure formation and the far future.

Let's first have a glimpse of modern chronology of universe.

The very early universe, started from the Planck epoch until the cosmic inflation, or the first picoseconds of cosmic time; this period is the domain of active theoretical research, currently beyond the grasp of experiments in particle physics.

The early universe, from the Quark epoch to the Photon epoch, or the first 380,000 years of cosmic time, when all the primary forces and elementary particles have emerged but the universe remains in the state of a plasma, followed by the "Dark Ages", from 380,000 years to about 150 million years during which the universe was transparent but no large-scale structures had yet formed;

The era of large-scale structure formation, incorporates stellar evolution, galaxy formation and evolution and the formation of galactic clusters and superclusters, from about 150 million years to present, and prospectively until about 100 billion years of cosmic time; the narrow disk of our galaxy began to form at about 5 billion years. The solar system formed at about 9.2 billion years, with the earliest traces of life on Earth emerging by about 9.8 billion years.

In the far future, after cessation of stellar formation with the diversified condition, this will decide the ultimate fate of the universe.

When we look the Vedic Chronology of cosmos it will be divided by **Dus Maha-Vidhya.**

It was only Adi-Parashakti who transformed herself by the String (Naad) of Mass in Singularity to the various forms of Energies.

The Devi Durga is the essence of universe, all the laws works because of Energy Adi-Shakti. Even in each phases of the universe whatever the forces evolve are because of energies. As per the theory ascertain energy is required to produce a force.

In further section, I will execute the role of ten secret wisdoms (Dus Maha-Vidhya) in various phases of the cosmos. These ten Maha-Vidhya are also the ten stages of phase transaction of universe from beginning to its existence.

The Dus Mahavidyas or the Ten Goddesses are actually ten aspects of the Shakti (Energy) or Devi the Divine Goddess. The Goddess Adi-Shakti the source, subject and the sustaining power of entire Creation (Kriti). **Prakriti,** this nature is realistic image of the Adi-Para Shakti. In Vedic, Prakriti (Pra means Pratham the very first and Kriti means the creation) so the very first creation that come in existence was Shiva elegant and feminine principle.

The ten aspect of Adi-Parashakti is the ten phases of cosmic growth is the nature (Prakriti) of Adi-Shakti and its great knowledge or wisdom is called Maha-Vidhya. This wisdom is the aspects of Devi Parvati, who represents a spectrum of feminine divinity.

In Chamunda Tantra we find the names of Dus Mahavidya as-

Kali Tara Mahavidya Shorashi Bhuvaneshwari

Bhairavi Chhinnamasta cha Vidya Dhumavati tatha

Vagala Sidhdhavidya cha Matangi Kamalatmika

Ete dus mahavidya sidhdhavidya prakirtita.

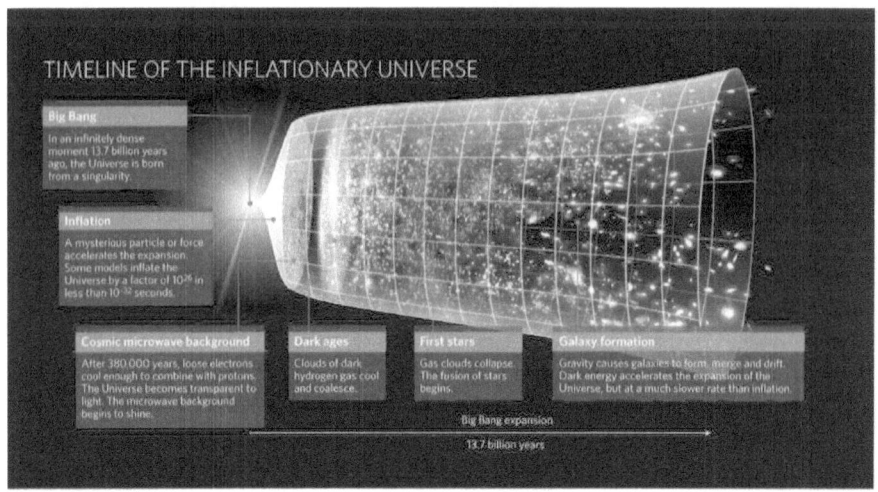

These Mahavidhyas are the phases of universe from Maha-Nadh (Big Bang) to its current position (Stithi).

The Big Bang (Maha-Naad) started by a string (Naad) of mass (Lingam) in the Singularity (Bindu or Yoni the field of energy), Yoni is the righteous word which is in misconception among some ethnic groups as a genital female organ, in fact it's the great field of energy associated with union and the generation.

It was mention in all Puranas; especially the details are given in Shiv-Puran and Markandeya Puran which state that there was only darkness before the origin of this universe.

That state was the state of ultimate darkness, the darkness was nothing but Goddess Maha-Kali apt a Non Dense Mass within her, non dense Mass is non other then Lord **Nishkal Shiv** as Maha-Kaal.

This Great Dark Age was before the beginning of big bang (Maha-Naad) and this is nothing but Goddess Maha Kali the "Ahim" (Singularity).

That's a great wisdom is not just a religious cult it's the supreme ancient science, which is experimented by the ancient scientists with their different observation, but ultimately the method remain the same as described long back in Vedas.

Those ancient scientists are called Seers or Rishish they are having strong calculation based on their experimental observation.

The only difference is modern scientists work in their Laboratories while ancient Scientists (Maha-Rishies) work in Forests and Himalayan region; they had dedicated their life to acquire the knowledge/wisdom, to accord the world with ultimate truth.

Now let's come to the point, we are talking about the first great wisdom (Maha-Vidhya) the **Kali**. So let's fall in a realistic story.

When we look the Big Bang model we find that before Big-Bang and after the inflation there was a Dark-Age, the period before the big bang was Maha-Kali the Singularity (Great Dark Age) and the age after the **Inflation** of Space is the divine principle of the same great energy Maha Kali as a Kali, This **Kali** was the first great wisdom (Maha Vidhya) because all the cosmological event are rooted from the Dark Age by the virtue of divine Goddess Kali.

Shital:- So you are revealing about the Maha Kali and Kali that means these Devi's are same and also what about Bhadra Kali or Kaal Ratri?

Vijay: There is only one Divine Energy which is manifested in various forms to establish the Law of nature and the same was proved by modern scientist by the law of energy which state that the energy can neither be created nor it can be destroyed but can be transformed to various forms.

The Adi Shakti is transforming herself to reciprocate the law of nature by its various forms.

Maha Kali, Kali and Bhadra-Kali, is the manifested forms of Singularity, She is recited as a single word of Beej-Mantra **Kreem.**

So the Maha-Kali is the Primordial Darkness, the Ultimate Darkness, the darkness which is the source of origin and also the end of everything and she takes the responsibilities of engulfing everything within her when there is imbalance in the system.

This is the phase of primordial Darkness from which with Anhad-Naad inflation emerged, this is how **Shri Vishnu** become extensively broad as a Space.

The Dark Age soon after Inflation was the source of all the objects of cosmos. Kali is regarded as an Adi Maha Vidhya (the primordial wisdom), she is everything whatever existing in this cosmos.

When we look the modern scientific model of Big-Bang we find various Phases of the universe like Inflation, Dark-Ages, the formation of first Star, Origin of Galaxies, and continuously growth of the universe. This model was presented by Modern Scientist.

The more advanced principles and theories was presented in Vedas by ancient Rishies, the Sadkas and our discussion is the proof of it, what I am narrating you is the ancient Vedic Science.

By going through Vedas I found nothing new in modern theories as everything remain same but only names of theories are changed.

But when we go's back into the time, the time of Vedanta which in reality cracks all our modern perception. In Devi Mahatamya which was endorsed and narrated long back 5000 years. The Vedic science is an interdisciplinary journal covering fundamentals, innovations and advance research. It is not just the science but an ecstatic universal principle.

Now if we compile modern scientific theories of *standard model of cosmology* with ancient universal Vedic principle of **Dus Mahavidhya** then all the secret of Big Bang model will breaks.

The dual natures of Divine Deities are manifested and unmanifested i.e. in form and formless. Their role is in every aspect of cosmos.

There were different phases of dark ages at different stages of universe, when that **Kali** manifested herself for creation, recreation, maintain or to destroy the evil forces of antimatter.

During the earlier course of our discussion, as narrated by Shri Brahma Ji to Deva Rishi Narad Ji, that chronic first Dark-Age before Big-

Bang is all about the story of transformation Nishkal Shiva (Energy) to Shakal Shiva.

So that was the first phase of Darkness which can be symbolize as Maha Kali, then there was inflation.

Nishkal is the unmanifisted form of Mass and Energy while Sakal is the manifested form of Naad and Bindu. From this Naad and Bindu, Space is emerged and from Space the Time. This is the sequence of universe.

Followed by Mass, Energy, Space and Time the universe came in existence.

5.15 Kali – The Dark Ages, an Era of Darkness That Exist before the Origin of Stars and Galaxies

This is the phase of the universe when the Darkness was enwrapped all around the space and Goddess Kali was creating a platform to universe to arrange the celestial bodies which she was supposed to originate.

The Dark Age or Kali is the mother of universe and she affectionate her entire creation and it is Fierce and Horrible for those Antimatter that was distractive for her creation.

This is the usually the character of motherhood, the mother is not allowing any distractive force to come near her loving creation.

Kali is the countenance surface holding everything within her she is the feminine energy meant to reproduce the creation.

So scientifically the Dark Age represent the period between the release of microwave radiation and the formation of first shining object. The Shining one is nothing but the first star (**Tara Devi**) which was transform from Dark- Energy.

The Phenomenology behind it was the event during Dark Age, that there was huge dark energy with abundance of dark-matter. The collapse of Dark-Matter which was caused and influenced by Dark-energy results in the origin of Shining object.

The Dark Matter of this Dark Age as we have already discussed in the story narrated by Shri Brahma Ji to Dev-Rishi Narad during the summon of **Rudras. Vedanta** Say clearly that there was a divine pageant between Adi-Shiv and Shiva (Kali) due to their Initial Spurt Goddess **Tara** (first Star) appeared.

Rig-Veda states that she is the primordial energy, she inherit whole cosmic deities and their power. She is the latent image for that period of latency, though she transformed in a manifest form in Patent Period.

It is clearly mention in **Nashadiya Sukta** of **Rig-Veda,** described as darkness is wrapped with in darkness.

Shital:-There are numerous myths about the Devi but now I understood scientifically and her fact of being of great significance.

Vijay:- The references of Dus-Mahavidhyas are from (Vamana-purana 30.3-9). The Devi is thought to assume these different incarnations in an attempt to maintain cosmic stability (Devi-mahatmya 11.38-50).

KALI- This "Devour and Emanation of Time". She is fierce in appearance but givers of ultimate prosperity to the cosmos we live in.

TARA- (meaning of Star in Sanskrit and in Hindi is Tara) She is the 1st primordial Star and the energy accountable for the formation of all the Stars of universe. The Goddess is worshiped as guide and protector and saviour the one who gives ultimate knowledge which gives salvation (also known as Neel Saraswati).

LALITA TRIPURA Sundari – **Shorashi** or **Tripura Sundari** – The Tripura Sundari is the goddess and her presence in all three worlds that means the formation of *Bhur (Planests),* **Bhuva (the gap between Stars and Planets) and Swaha (Galaxies)** also the most beautiful among all, literally means the formation of Planets, Galaxies and Heaven (the world beyond our galaxy) this is the third phase of Big bang Model.

Also known as shodashi, she is the one who is "most beautiful in the Three Worlds" (Supreme Deity Srikula systems); the "Tantric Parvati" or the "Moksha Mukta".

BHUVANESHWARI- the Goddess as World Mother, or Whose Body is the Cosmo. – Bhuven (World) + Ishwari (Goddess) is the expanded universe, is the Goddess of the entire universe.

BHAIRAVI- The fierce warrior Goddess.

CHINNAMASTA - The self-decapitated Goddess, who holds her neck in her hands.

DHUMAVATI- The Widow Goddess or the Goddess of death.

BAGALAMUKHI- The Goddess Who Paralyzes Enemies.

MATANGI- the Prime Minister of Lalita (in Srikula systems); the "Tantric Saraswati".

KAMALA-The Lotus Goddess; the "Tantric Lakshmi".

Origin of Sun and the Solar Deities.

Shital:- You have shared the secrets of Vedic cosmology, but I want to know how our Galaxy and Solar System begins, Is there any scientific description in Vedas, Puranas or in Upnishads?

Vijay:- Off course, but before understanding the formation or the beginning of Solar system, you should aware about the Adi-Shakti in the form of cosmic matrix, these cosmic matrix is called Aditi. From **Aditi** the solar deities got originated. Mata Aditi is having double role same as we found in Indian movies.

Shital:- Double role, in which way?

Vijay:- Mother Aditi before she appeared as a daughter of Daksha Prajapathi, She is the cosmic matrix. The birth of our galaxy begins by the origin of Sun God.

To understand the origin of sun and the other solar deities we have to go ones again in the epoch of **Radiation era** of **Anhal-Stambh.**

You remember we have discussed in earlier narration about the dialogue between Narad Rishi and Brahma ji.

This was the incident when Shri Narad Ji asked his father about the origin of the universe and the percept of **Shivatatva**.

Shital:- Yes I do remember, Shri Brahma ji narrated very interesting and remarkable phenomena of late universe, When the last cosmic scale was on its peak, then the universe was annihilated by Lord Rudra, interns there was nothing left as a remnant, entire (चराचर) world achieved the state of Singularity and the Darkness (Dark Energy Goddess **Kali**) distributed throughout and that the congenial span.

The ultimate fate of the universe is a topic in physical cosmology, whose theoretical restrictions allow possible scenarios for the evolution and ultimate fate of the universe, to be described and evaluated. Based on available observational evidence, deciding the fate and evolution of the universe have now become valid cosmological questions, being beyond the mostly untestable constraints of mythological or theological beliefs. Many possible futures have been predicted by different scientific hypotheses, including that the universe might have existed for a finite and infinite duration and explained the manner and circumstances of its beginning.

This was achieved something like; on a macroscopic scale, as matter aggregates into larger and larger bodies until they collapse in upon themselves, the nothingness of compressed uniformity is approached. This tension between infinite cycles of compressed and expanded uniformity endows the universe with its particular form. Nothingness theory is the exploration of how the universe manifest in scientific, philosophical, and theological principles.

Then from Nishkal turn to Sakal, thereafter redistributes mass (Shivatva) evenly approaches expanded inflation of Maha Vishnu from Aim the Adi-Shakti and Shiv.

The appearance of Anhal Stambh after the inflation of Shri Hari Vishnu as a Space and the beginning of Time by the appearance of Lord Brahma from the naval cord of Shri Vishnu.

The Lord Brahma and Vishnu could not able to found the end of the **Stambha**, it was infinite object with infinite distance. I remember this fact of our discussion. Something more secret was hidden over there during that epoch of radiation?

Vijay:- Obviously When Brahma and Vishnu understood the cognition of that **Stambha**, they found the secret science of **Gayatri** and **Maha Mirutyunjay** Mantra.

So now we will discuss about the **Gayatri Mantra** and the origin of Sun our master solar deity. We will recite the same Mantra before the beginning of our discussion.

ॐ भूर्भुव: स्व: तत्सवितुर्वरेन्यं । भर्गो देवस्य धीमहि, धीयो यो न: प्रचोदयात् ॥

Om Bhur Bhuvaḥ Swaḥ

Tat-Savitur Varenyaṃ

Bhargo Devasya Dhimahi

Dhiyo Yonaha Prachodayat

Word meaning: **Om:** The primeval sound; **Bhur:** *BHOO* (भू) the heavenly and celestial body like earth/physical realm; **Bhuvah:** the life force/the mental realm and the atmosphere **Swah:** Swarga (स्वर्ग) the soul/spiritual realm/heaven; **Tat:** That (God); **Savitur:** the Sun, Creator (source of all life); **Vareñyam:** adore; **Bhargo:** effulgence (divine light); **Devasya:** supreme Lord; **Dhīmahi:** meditate; **Dhiyo:** the intellect; **Yo:** May this light; **Nah:** our; **Prachodayāt:** illumine/inspire.

The *Chhandogya Upanishad* explains the significance of the first line. It tells us that once Prajapati Brahma, the Lord of the Universe, contemplated the nature of the three worlds (Anahal Stambh) earth, sky, and heaven—and through intense concentration he was able to discover the essential guiding force of each: *Agni* (fire) governed the earth; *Vayu* (the vital force) governed the sky; and *Aditya* (the sun) governed the vault of heaven.

The next two lines of the Gayatri mantra venerate the concept of **Savitur**- the solar deity of light, energy, purity, transcendence, illumination, and compassion (the sun shines for all). It is:

tat savitur varenyam

bhargo devasya dhimahi

This translates as: "We recall within ourselves and meditate upon that wondrous Spirit of the celestial Solar Being." In simple it can be understood as: Tat: That Savitur: is the Sun and also the jyothirmay Lingum the source of light. So it's clear that the source of light after radiation era was the solar deity **Savitur**.

It describes the *bhargah* (the solar spirit), who is the essence of *Savitri* (the solar being), who is yet the inner identity of *Surya* (the sun). The Gayatri as a prayer is a petition to *tat* **Savitur** (that Sun) which is the infinite light of pure consciousness.

Now, the light which shines above in heaven, pervading all the spaces, pervading everywhere, both below and in the farthest reaches of the worlds—this indeed is that same light which shines within man.

—(Chhandogya Upanishad 3.13.7)

Vijay:- This is what Shri Brahma and Shri Hari Vishnu saw in Anahal-Stamb, the secret of Gayatri Mantra which says that lord Shiva himself is **Savitur** *after big-bang (Anhad Naad) was appeared in the Anahal Stambh with primordial sound OM. Not only this Gayatri Mantra also tells, about the source of light;*

Om: with Primordial sound; **Bhur**: the physical body/physical realm the element like earth; **Bhuvah**: the life force/the atmosphere (Gasous like Hydrogen and Helium i.e.)**Suvah**: the floating heaven; formed **Tat**: That (God); **Savitr**: the Sun, Creator (source of all life); **Varenyam**: adore; **Bhargo**: effulgence (divine light); **Devasya**: supreme Lord; **Dhīmahi**: meditate; **Dhiyo**: the intellect; **Yo**: May this light; **Nah**: our; **Prachodayāt**: illumine/inspire.

So it can be said like, with primordial sound OM, earth, sky and heaven or mass, space and time together formed the Sun god **Savitir** in the **Anahal Stamba** with the help of matrix of **Savitri** (Aditi, the infinity of Anahal Stambha or Jyothirmay Lingum). Gayatri Mantra is also called Ved Mata.

Now the question is that the **Savitr / Savitur** and the Sun which shines our planets is same or difference.

Answer for this is, there are numerous shining objects i.e. sun like stars in the universe but the primordial spirit of Savitur remains the same in all Sun's. That means **Savitur** is the one source of light, for us and our planet, it's our Sun (Surye Deva) and in the beginning of the Universe it was Anhal-Stamba (the primordial fireball).

5.16 The Supernova "Origin of Our Sun"

Vijay:- There is an interesting Vedic story, which is the same doctrine of modern science of supernova, In the essence a supernova is a violent stellar explosion it pretend that the Sun (earlier sun) went supernova.

Shital:- I am keen to know such description given in Puranas, it's so scientific and spiritual too.

One thing I understood is that there is always someone (Almighty) behind each and every phenomenon of nature. This can be called the theory of cause and effect, nothing happens by chance or outside the Universal Laws.

The **Savitr** enlighten the entire world after the radiation epoch of early universe until the birth of Surya (Sun god) from Aditi and Kashyap.

What I understood with Savitr is, Savitr is the god of light distinct from the Sun. Now I am curious to know the phenophase and the relation of **Savitr** and Surya (our solar deity).

Vijay:- Absolutely, Savitr is the light leftover after Jyothirmay Lingum or the Anhal-Stambh, then that primordial light shine with the origin of Tara Devi. It's the celestial light of very first star of the universe. Same light is the part of all numerous stars.

Now we can call in modern language **Savitr or Savitur** the **Photon** which is an elementary particle of the electromagnetic radiation of light, and the force carrier (even when static via virtual photons). The

Savitir an ordinary photon has zero rest mass and is always moving at the speed of light. Photons are having both wave and particle nature.

The photon is accounted for the frequency dependence of light's energy, and explained the ability of matter and electromagnetic radiation to be in thermal equilibrium.

The **Jyothirmay Lingum** or **Anhal Stambha** was in the state of Plasma in the beginning of universe, the Space Time odyssey begun from the Quark epoch to the Photon epoch, till the first 380,000 years of cosmic time scale as per modern science and the same Photon epoch is narrated as epoch of Savitur. The universe was followed by the "Dark Ages (**Goddess Kali**) and then the appearance of 1st shining object as **Tara Devi** (very first star of early universe) is the god **Savitur**.

Shital:- Are any more traits from Vedanta is available in support of Savitr as a photon?

Vijay:- Exactly, the Rig Veda also states the same execution, we will analyse the hymn of Rig Veda.

> *a kṛṣṇena rajasa vartamano nivesayann amṛtam martyam ca |*
> *hiraṇyayena savita rathena devo yati bhuvanani pasyan ||1.035.02*

ā | kṛṣṇena- Darkness | rajasa- Rajas character | vartamānaḥ- Now| ni-veśayan- Insertion, expansion| amṛtam- immortality | martyam- mortalise | ca- with | hiraṇyayena-Golden rays | savita-Solar deity | rathena-Chariot or travelling | ā | devaḥ-Deity | yati-this | bhuvanāni-universe | paśyan-seen or Sheen

This means; O Deity Savitr, you are the lord of light now (vartamanaḥ) present and expanding the immortalize (amṛtam) and mortalize. Your rays (Hiranyayena) are pervading and transporting throughout the space, removing away the darkness (kṛṣṇena) from the universe (Bhuvanani).

Savitr deity, in the form of photons that are travelling and roaming as the hymn says (Rathen) means moving like a chariot, these photons take off and remove darkness.The term photon (meaning "visible-light particle") and the cause of these visible light is Savita Dave. We can simply say the photon is the fundamental particle of visible light emerges in the form of God Savitr.

The earliest photons probably appeared about fifteen billion years ago, during the Big Bang or **Anhad Naad.**

Shital:- Gee' how exquisitely you have explained Savitr and his relation with photons. I know the modern definition of Photon is *"A photon describes the particle properties of an electromagnetic wave instead of the overall wave itself. In other words, we can picture an electromagnetic wave as being made up of individual particles called photons. Both representations are correct and reciprocal views of electromagnetic waves. For example, light exhibits wave properties under conditions of refraction or interference. Particle properties are exhibited under conditions of emission or absorption of light."*

Now more I want to know is that why these photons are mass less?

Vijay:- This Shakal-Shiv is the total mass of universe and his Nishkal Swaroopa (form) is mass less, the radiating primordial fireball appeared after Anhad Naad. Brahma ji Said to Narad Rishi about the same.

Brahma Ji Said:- Hey Devrishi, In Anahal Stambha (Primordial fireball) there was a divine light but not the Sun. That luscious light is of Shiva emerging as Savitri.

Vijay:- It was described by modern science that after Big Bang there was a light but nowhere was the object like Sun. We have already discussed this in the beginning of our discussion that CMB is the Pashupath of lord Shiva.

The same has been referred by Modern science as **cosmic microwave background (CMB which is the truly the Anhal Stamba)** which is the thermal radiation left over from the time of recombination in Big Bang cosmology. In older literature, the CMB is also variously known as cosmic microwave background radiation (CMBR) or "relic radiation". The CMB is a cosmic background radiation that is fundamental to observational cosmology because it is the oldest light in the universe, dating to the epoch of recombination. This glow is strongest in the microwave region of the radio spectrum. The accidental discovery of the CMB in 1964 by American radio astronomers Arno Penzias and Robert Wilson was the culmination of work initiated in the 1940s, and earned the discoverers the 1978 Nobel Prize.

Vedic and Puranic set of statements, I have constructed to describe a set of facts which clarifies the causes, context, and consequences of those facts involve in the origin of theuniverse. The cosmic microwave background radiation (CMBR) details are mentioned in Shiv Puraan, which is called the Pasupath, the faint glow left from Jyothirmay Lingum or the Anhal Stamba after its disappearance. The only difference is of terminology of Sanskrit text and modern science in English.

Photons of Savitr deity is free to sail around the world unimpeded and these photons enabled to continue their voyage, without any hindrance. The decoupled photons **(Savitr / Savitur)** originated from the great primordial fireball (Anhal Stamha) of Lord Shiva were free-streaming across the universe. This fact was observed by Shri Hari Vishnu and Brahma Ji.

This was the anecdotal phase of **Jyothirmay-Lingum** and the next phase of the Universe was Dark Age of Goddess **Kali** which we already discussed in our earlier conversation.

Then we know a few hundred million years later, new sources of energetic photons (Savitr) appeared from first star Goddess **Tara Devi**, Stripping hydrogen **(Sunand)** atom returning them to their ionized state, ultimately allowing light to travel easily through the intergalactic medium. After this era the reionization was completed and the universe was fully in transparent state once again.

5.17 The First Supernova

The Maha-Vidhya, Chhinnamasta Devi
Habitable epoch of the Universe

Vijay:- Shital, you might remember our earlier discussion about Dus Maha-Vidhya now the time comes that I should narrate you about the story of Devi- **Chhinnamasta**. The Goddess created the habitable zone and the ideal circumstances for early cosmos. Remember "Dus-Maha Vidyas" have different roles applicable on each and every aspect of life, objects and inhabitants.

This is the scared knowledge about the early supernova and the origin of our Solar Deity Surya along with other amity to encore different galaxies.

Now, If I ask you about **Devi Chinnamasta,** what would you say about her?

Shital:- I can say only that she is one among Dus-Maha Vidyas and one morph of Devi Parvati or Adi-Shakti.

I know a legend, attributed to the *Narada--Pancharatra* and the tale is as:

Narad ji Said:- Once, Devi Parvati with her two companions was wandering in Himalayan region, Thereafter Parvati started bathing in river Mandakini with her two attendants Dakini and Varnini (also called Jaya and Vijaya) after long play and pleasure in water both of her friend become extremely hungry and beg for food.

Though Devi Parvati initially promises to give them food once they return home and she was again busy in taking bath, by the time both of her friends was starving and both of them again requested to return home so that they can have food, on this Devi **Parvati** looks in and around but couldn't find anything to eat.

To appease the hunger, the merciful goddess eroded her head, the fountain of three stream of blood from her neck flows in three directions; one stream in Jaya's mouth, second in Vijaya's mouth and the third stream in sliced head of Parvati's own.

The mother Devi Parvati again pair her head and came back to home with her attendant jaya and vijaya.

This is the story I know about Goddess Parvati as she chop her head so called as **Chhinnamasta,** I want to know more fact about this if you can?

Vijay:- Absolutely, this is the one aspect of Adi-Shakti mata Parvati she is the retentive of all ten cosmic powers of universe. **Chhinnamasta** is the one among these powers. The **Chhinnamasta** in her exhaustive form is the cosmic power.

She symbolises both aspects of Devi: a life-giver and a life-taker, Chhinnamasta is described as a red Goddess; she shines even with more energy that the millions of suns radiated over their lifetime. She is usually depicted as red in complexion. Her red complexion indicates the same as any star before supernova turns into red giant.

One invocation recited for her sacrifice and the fulcrum of the sacrifice, with the severed head treated as an offering.

This paradox signifies the entire sacrificial process of Devi-Chhinnamasta, and thus the cycle of creation, dissolution, and re-creation as the act of her dalliance.

The story you narrated also have science of fiction which now I shall deliver you.

Shital:- Oh, If it is so, I want to know that now?

Vijay:- We have discussed almost all the important epochs of our early universe but an important epoch is called early universe supernova epoch.

When we look at the night sky, we see the uncountable stars and all these stars are the linage of **Tara Devi** and the cosmic power which holds the galaxies is Devi **Tripura-Sundary**, the infinite universe bolster by numinous Devi **Bhuvaneshwari,** the science of Adi-Shakti was already discussed by us.

Our own origin keeps a great cosmic question in front of us, but literally from these facts and events of the cosmos we can infer lots. Even in our today's Solar System, we cannot know everything that transpired to bring us about into our present state just as like the rest of the Universe, when we view things today, all we see are the survivors, with the remaining details lost in the past history of time. However, the Universe provides us with enough clues that we can draw many reasonable, robust conclusions about a whole swath of ideas, including our own existence.

Now, I will elaborate how, the planetary system came in existence and the universe became habitable. Beyond our own solar system, there are numerous planets and stars. So far, we have discovered thousands of planetary systems orbiting other stars.

Billions of stars in our galaxy are thought to have planets of their own. Milky Way is one of perhaps 100 billion galaxies in the universe. This work was accomplished by the cosmic power called Mata **Chhinnamasta**, She sacrifice her to make the universe a habitable zone for all beings. The secret science I will reveal you about Devi **Chhinnamasta** is the insight of mystery and signifies spiritual truth regarding the fulfilment of worldly desires.

When the universe was filled with Population III stars (the classification given by modern **astronomers**). Those first generation stars were likely very large, about a hundred times the mass of our Sun. These are the population of stars composed entirely of primordial gas – **hydrogen**, **helium** and very small amounts of lithium and beryllium.

They all are being the progeny of Rudra-Tar and Tara-Devi and operated by Tara-Devi with the association of **Sunand** and **Nandan** as a Rudra, hence population III stars are called as **Rudra-Star**.

These the very first generation, Population III stars are completely devoid of metals, the metals observed in Population II stars and initiate the gradual increase in metallicity across subsequent generations of stars. These second generation stars have traces of elements such as carbon, but universe is still lack of heavier elements like Iron.

These all phase transaction was accomplished by the energy of Devi **Chhinnamasta**. She is the energy of all stars and sacrifices herself by the evidence of supernova.

Now we will analyse the feature and function of **Devi Chhinnamasta**, She is usually depicted in red complexion which symbolize the color of Star before they undergo supernova, the Stars with less mass become red super giants.

Supernova: Five Stages involve in the death of a Star and the role of Goddess **Chhinnamasta.**

1. Just before explosion a red super-giant star approaches the end of its life. There is no more fuel to burn and make it shine. Soon its massive dense core is bound to collapse under its own weight.

    Now you will understand, both modern and Vedic scientific phenomena will remain the same, Devi **Chhinnamasta** color is Red before she chops her head. There was nothing to eat, thus she collapse her own head with self potential.

2. The first light flash the core collapses and sends a shock wave out. For a few hours the shock compresses and heats the envelope, thus producing a very bright flash of light from the inside of the star.

    Producing flash of light and sends shock way out is when she refer as a cosmic power her in the form of sun or star, same way as she produces the stream of blood from her slashed neck in the aspect of **Chhinnamasta in her** Parvati Swaroopa.

3. The flash has gone after hitting the surface at 50 million km/h the shock blows the star apart. The core turns into a neutron star, a compact atomic nucleus with the mass of the Sun but 10 km in size.

Then that flash stream pacifies the surrounded space to form a new star similarly she pacify her friends in her Parvati- **Chhinnamasta** swaroopa.

4. With proper Supernova, the hot glowing surface expands quickly making the fireball brighter again. In a few days it will be 10x the size of the original star and will be discovered by supernova hunters.

5. A long time after the remains of the former stars are spread over light years of space. They keep floating quickly, sweeping up interstellar gas here and there, leaving a faint beautiful glow behind.

Though the head was cut off but stellar evolution stages began of massive stars in both of her form or we can say Goddess **Chhinnamasta** is the root cause of life and death of Stars.

These all was approached by the energy, the power of Goddess **Chhinnamasta**. She represents death, temporality, and destruction as well as life, immortality and recreation of Stars.

Performing the play, Parvati Devi with her attendant Jaya and Vijaya is just to proclaim her cosmic realm and capacity.

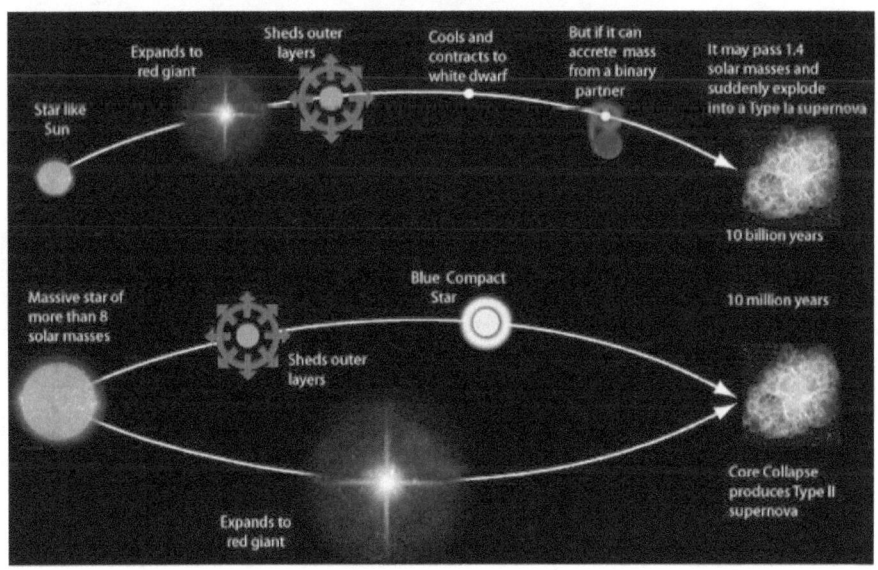

Devi Chhinnamasta performs & expedites the Stages involve in the death (Supernova) of a Star

Chhinnamasta is a morph of radical transformation, a great *Yogini*. She conveys the universal message that for sustainability of life in any form whether being or celestial, destruction and sacrifice are necessary for the continuity of creation. The goddess symbolises *Pralaya* (cosmic dissolution), where she swallows herself and makes the way for new creation, thus conveying the idea of transformation.

This is how the present generation of stars originated and these stars can be spotted in today's sky. From these stars we got abundance of heavy elements.

We have discussed only about the first generation population III Rudra-stars. Namely there are 3 generation of Stars. The Devi Bhuvneeshwari as a goddess of galaxies is the energy associated with population III stars.

Shital:- If we categorize the generation of stars may fall as under:

1. Population III Stars as 1st generation.

2. Population II Stars as 2nd generation.

3. Population I Stars as 3rd generation.

But as you said Population III Stars as 1st generation are Rudra Stars then what about other two categories and how the universe reaches at this stage.

Vijay:- Astronomers classify stars, they normally lump them into three categories, creatively named Population I, II, and III stars. Population II Stars as 2nd generation started by the beginning of 1st generation of deities.

Some Stars fall under Rudra Categories while some are Vishnu and Brahma. In fact all three generation of stars share the interchangeable characteristics of Rudra as a matter, Vishnu as space and Brahma as time in association of energy.We already discussed about the origin of identical Rudra's who become the 1st generation of Stars in the form of Agni and the solar deities like Sun or Savitr (Savitur) of distinct galaxy (it's not today's sun of our galaxy).

The objects which shine in sky may be the aspect of Brahma, Vishnu or Shiva having their Gunas in different ratio.

Population II Stars as 2nd generation are Sapt-Rishis as a Manasputras of Brahma hence they are the progeny of 1st generation of Stars. We can easily identify this group of seven stars in the night sky.

The transition phase of each generation leads by supernovae and finally the universe got abundance of elements by those supernova stars.

Same way the 2nd and 3rd generation Stars supernova remnants also produced the Vasus which are the mainly the elements specially presents in Planets.

The transition phase of each generation leads by supernovae and finally universe got abundance of elements by those supernova stars.

Shital:- It's now explicit representation about the 1st generation of stars but what about 2nd generation the Manas Putras as Sapta-Rishi's, Please let me know?

Vijay:- To understand this, let's go reversal in the anecdotal story of Brahmas creation and in the interaction of Shri Brahmaji and Devrishi Narad. You might remember that after the birth of four Brilliant Son's Sanak, Sanandan, Sanat and Kumar.

On their refusal of helping Brahamaji in the process of creation, Lord Rudra appeared and created numerous identical Rudra Ganas which become the 1st generation of celestial stars. There are the seven Rishis who are called as Saptarishi extolled in Vedas and *Upanisads*.

Come let's move to be the part of Shri Brahamji and Devrishi Narad Ji discussion.

Brahma Ji Said:- Hey Narad, I am the creator of time. Despite all these creations, I was not satisfied. I created sage Marichi from my eyes, Sage Bhrigu from my heart, Sage Angira from my head, Sage Pulaha from my Vyan Vayu, Sage Pulatsya from my Udan Vayu, Sage Vashishth from my Saman Vayu, Sage Kratu from my Apan Vayu, Sage Atri from my ear,

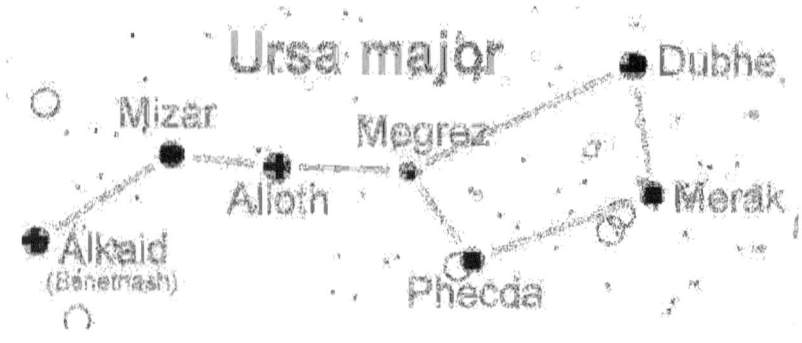

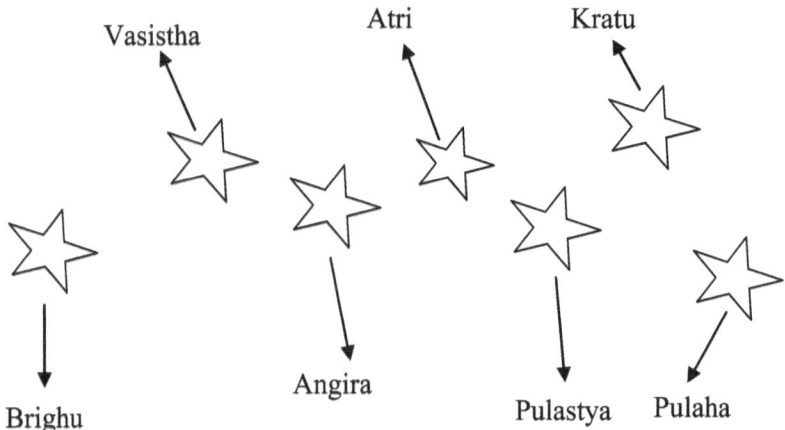

Population II Stars- 2nd generation Stars are Sapt-Rishis as Ursa Major

Vijay:- Brahmaji was narrating the scared science of his creation of Population II Stars as 2nd generation. He said **Marichi** was created from his eyes.

Shital:- So what exactly is **Marichi**?

Vijay:- As part of the creation of this world, Lord Brahma created his sons 'Manasaputras' and 'Prajapatis' to whom he can delegate the responsibility of different tasks of creation. Marichi means the ray of light; we know the source of the ray of lights is Stars. Hence Sage Marichi is the ray of lights and we can spot this Sage star in Usra Major Constellation. Brahmaji the master of time created **Marichi**, the processes was followed by his hot spot sight; he transformed the light and energy and made the radical structure called Marichi.

Then after some time Maharishi **Bhrigu** was created by Brahma Ji, Maharishi Bhrigu is a great scientist and he is the celestial being in appearance, he is the star and he was blessed by the attribute to adopt human body when needed.

Angira (Western Name Alioth ε Uma) Brahmaji then created Angira from his head– with his Ikchha (wish) and buddhi (divine intellect). Brahmaji granted him the celestial divine glow and inner spiritual capacity of power, wisdom and divinity as the 3rd star among Sapt Rishi Mandal (Ursa Major). Through his Tejas, i.e. divine glow, Angira enlightens the whole universe.

When we deeply study the duties, functionality and origin of Angira, We found the **Angira** as a 2nd generation star. Angira literally means in Sanskrit the one who posses fire matter, in other words stars are also called Angira and the structure is associated with matter in its core with abundance of hydrogen and helium.

Angirasa is a Rishi (Sage) who later, along with Sage Atharvana, has credited and formulated the fourth Veda called Atharva Veda. Alongside, the episodes of Angirasa have also been mentioned in the Rig, Sama, and Yajur Vedas. In the scriptures he has also been referred to as one of the seven sages or Saptarshis of the first Manvantara with others. His wife's name was Surupa. He had three sons namely Samvartana, Utatya, and Brihaspati.

Angirasa is closely associated with Agni by means of Hydrogen and Helium. Angirasa is identified with Agni:

यदंग दाषुशे त्वं, अग्रे भद्रं करिष्यसि।
तदैतद् सत्यमङ्गिरः॥ - Rig-Veda

Same way Lord Prajapathi Brahma created other Star' respectively Pulaha from his Vyan Vayu, Sage Pulatsya from his Udan Vayu, Sage Vashishth from his Saman Vayu, Sage Kratu from Apan Vayu, Sage Atri from his ear.

Shital:- You stated that the 2nd generation stars named Angiras, Pulaha, Pulatsya, Vashisht, Kratu & Atri are formed from gas. The modern Astronomers also spotted that *2nd generation stars* made up of the remnants of **gas** dust, this is evidently true?

Vijay:- Absolutely true, the formation of Sage Pulaha, Sage Pulatsya, Sage **Vashishth** and Sage Kratu from his various air forces, Vayu means the gas so the formation of these Sage's birth begins by Brahmaji's cosmic gas accretion controls and result in the formation of star, relationship between stellar mass, metallicity, and star formation rate (the so-called fundamental metallicity relationship).

After the big bang or Anhad Naad, the only elements in the universe were hydrogen, helium and trace amounts of lithium. There was no carbon, oxygen or iron, because these elements are only formed when stars undergo fusion in their cores by Second and third generation of stars.

About 100 to 300 million years after the big bang, the first stars began to appear, with the aspect of Mahavidhya Tara Devi. Tara Devi the 1st star of first generation stars were likely very large, about a hundred times the mass of our Sun. These 1st generation stars because of their big size, they had short lives which ended as supernovae. From the remnants of those stars, a new generation of stars was generated by Lord Brahma Ji. These second generation stars would have traces of elements such as carbon, but still lack heavier elements such as iron.

In astronomy, all elements except hydrogen and helium are referred to as "metals." For this reason, a measure of the amount of other elements a star contains is known as its metallicity. One way to define the metallicity of a star is simply as the fraction of a star's mass which is not hydrogen or helium. For the Sun, this number is $Z = 0.02$, which means that about 2% of the Sun's mass is "metal". Another way to express the metallicity of a star is by its ratio of Iron to Hydrogen, known as [Fe/H]. This is given on a logarithmic scale relative to the ratio of our Sun. So the [Fe/H] of our Sun is zero. Stars with lower metallicity will have negative [Fe/H] values, and ones with higher metallicity have positive values.

Modern ways of understanding for the origin and evolution of solar systems and galaxies is still very limited, it's based on the observation and research and it's not possible to go back in time to know to find what exactly happened. Our current observation may wrong for the past evidences. Pragmatically, the Vedic way in association with modern scientific phenomena may give some clear picture because Vedas are the archaic trait appraisal.

Shital:- Very well described, it's amazing to know Vedic knowledge in such a way, we even can't imagine how scientific was our Vedic period. Now I want to understand how this universe operates, as you told that the universe at this stage achieved habitable zone.

Now I understand nothing is autogenesis there was a big history and I remember you told in early narration about the split of Adi-Shakti from Shiva (energy from mass).

Vijay:- Once again coming in the story of universal binding force as narrated by Shri Brahma Ji to Narad Rishi.

But in realistic Mass and Energy can never separated, In Vedas and In physics, **mass–energy equivalence** states that anything having mass has an equivalent amount of energy and vice versa, with these fundamental quantities directly relating to one another by Albert Einstein's famous formula:

$$E = mc^2$$

Most people have heard about equation given by German-born physicist Albert Einstein's theory of special relativity "$E = mc^2$". Some know that Albert Einstein is the one who developed the mass-energy equivalence.

But I wonder how many actually understand how to use it or what it means. I have known about $E = mc^2$ for most of my life, but never really understood it beyond the mind altering thought that matter is energy and energy is matter. It is hard to move much past this simple point, because most scientists, engineers, and philosophers have trouble actually defining mass or energy. So what does $E = mc^2$ mean?

It's an equation defining about the relationship of Shiv and Adi-Shakti, the association of Rudras and their consort, the link between Shri Narayan and Maha Lakshmi, the cognation between minor deities (Demigods) and their consorts in their celestial and transcendental metaphysics form.

Same law is applicable for each aspect of the cosmos as a universal binding force.

Energy I'll define in an unsatisfactory but seemingly the only way possible, "That which is conserved and can be transferred as heat, work, radiation, or matter."

$$E = mc^2$$

Were E is an energy Adi-Shakti, m is mass the Rudra and C is the speed.

Now, let's move on to $E = mc^2$. It might be thought of as, if we took a piece of matter and made it "disappear", that we would get mc^2 amount of energy released, but does that ever really happen?

Do we ever just make a chunk of matter disappear? No, not really. So, let us instead think about an atom and what it is made up of: protons, neutrons, and electrons. If we measured each of these things – called subatomic particles – independently, we would find they each have their own mass (mp is the mass of the proton, mn the neutron, and me the electron). Now, what we find is that when we form a hydrogen atom from a proton is that the mass of the hydrogen is not equal to the mass of the sum of its parts. There is now a mass defect. So, what happened to that mass? It was converted to energy, meaning that forming a hydrogen atom from subatomic particles would release energy! Thus, the mass defect is exactly related to the change in energy by Einstein's equation: $E = mc^2$.

We can also think about this process in reverse. If we imagine a hydrogen atom in a box and we split it into a proton and an electron, we would end up with more mass. Therefore, assuming we didn't add any particles with mass to the box, we know that mass must have come from putting energy into the box. This energy is the amount of energy it would take to split a hydrogen atom, also called the binding energy.

Now we will take a same note on Shiv and Shiva, the Adi-Shakti seperated from Shiv and transformed in various forms to give a proper cosmic structure to the universe. The amount of energy transformed in any form having definite amount of mass.

We have discussed about six out of ten cosmic powers (wisdom) of Adipara-Shakti. In all her form there is Rudra (Shiv) as a consort. The concept of equilibrium can be understood more specific by way of Lord Shiv and Shiva as below.

Dev Rishi Narad asked:- O father of creature (Prajapati) I want to know how the universe is in operation and who ruled this cosmos. You have described the secret from beginning of the universe to the establishment of celestial bodies. I understood that it's Shiv and Shiva who are universal binding force while you are the creator and the celestial deities in the space are sustained by Lord Narayana. These are the key responsibility performed by you four.

Brahmaji Said:- Hey Narad, You may remember I told you earlier that, when I was unable to brought desirous growth of the universe I have started penance of Lord Shiva, then universe achieved the state of celestial population and habitable by Rudra and Adhi-Shakti in their manifold aspect. I started penance to please Lord Narayan.

Hey Narad, though Lord Shiv and Shiva are the cause of everything but they do not indulge in the bondage of this universe, Parbrahm do everything but remain free from any effect. *Shiva* (Shiva is the common word for Shiv and Shakti which can be called a Singularity) means, the Braham "the one who is eternally pure" or "the one who is not affected by three Gunas. Shiva is *Aadi* and *Anant.*

Narad ji asked:- Hey Pita, I want to know about the *anecdotal* story of Prajapati because If you are the creator, how my brother Daksha got the designation of Prajapati?

Brahma Ji said:-On my request for the proper creation, growth and functionality of the Universe, Parbrahm Shiva (Par is Adi-Parashakti Shiva and Brahm is Shiv) is separated into *Purusha* and *Prakriti.* Purush is the etymon seed of creation and Prakriti is the nature or Yoni the abode or source of creation.

Shiv appeared in the form of Rudra in every aspect of creation while Adhi Shakti became Sati, Prakriti, Aditi and she is in every aspect of nature, she pervaded into the various cosmic power. Lord Shiv is the Par-brahman and Parmeswari (Uma) is his supreme Adipara-Shakti (Moola Prakriti). They are indeed one, but for the purpose of creation, they apparent as two as – Uma (Prakriti) and Shiva (Purusha).

Hey Narad, even Lord Vishnu Shri Narayan as a preserver, protector and sustainer is the aspect of Yoni or nature; I am the seed in another form. We are differentiated only on the base of Karma or Guna, but we are originally the one, structurally world can differentiate us, but basically we are one.

So Devrishi, you remember the Great Lord Rudra Shiv transformed himself into many identical Rudra's on my request and they all rule the universe by deferent mean.

Same way the processes of creation can only achieved by the grace of Adi-Parashakti, when Adi Shakti pervaded throughout the universe by means of cosmic energy.

So the Adi Shakti are Dus-Mahavidhya's, the Durga, Laxmi, Saraswati, she is Aditi, She is everything what we perceive and Rudra's are all what we can see. Entire entity is the transformation of Shiv and Shiva.

Devrishi Narad asked:- Hey Pita (father) as you said Lord Rudra appeared in various forms in different Kalpas, now I want to know more about **Adipara-Shakti**, it's clear to me that everything emerged from Braham and the world is entity of manifested Braham and what is unmanifest is also the Braham.

You have already narrated how everything emerged from Nishkal Shiva or Aimkari Para Shakti (the only cause). I want to know more about the energy transformation by the virtue of Vedas.

Brahma ji said:- It's supreme divinity of Ardhnarishwara who taught the world about the significance of sacrifice.

Naradji:- Sacrifice, what does it mean Parampita?

Brahmaji:- Yes sacrifice, when I was supposed to perish in very early phase of Kalpas, It's Lord Ardhnarishwara who appeared and made me perceived. For the welfare of world, **Ardhnarishwara** split into shiv and shiva and blessed the creation.

Narad Ji asked:- Hey Brahmdev, I also want to understand as Vedas says Aditi is the mother of all Deities, even Prajapati-Daksha mother is Aditi but again in turn, the Puranas has described Aditi as the daughter of Daksha Prajapati, what is this dualism?

Brahma Ji Said:- Since the processes of creation was so long we cannot put all the events together, before the birth of Aditi as a daughter of Daksha, She is the matripotestal cosmic matrix of the universe.

Vijay:- Explaining to Devrishi Narad, Brahmaji started reciting the secret of Vak Suktam (Suktam means formulae) of Rig Veda. Literally vak suktam is also called Devi Suktam.

This is the science when Adhi Shakti defines her presence in warp & weft of each and every aspect. She herself is extoolitie, since the beginning of the universe is really just the start of the illusion of Devi Maha-Maya, *that seems all matter is apart from energy but truly both are the two sides of same coin.*

It's the plan of Almighty; we can't get rid of a creator and the self for the any reasons. It exceed all current laws of science since lord shiva himself say's that I am corpse without energy (शक्ति के बिना शव के समान हैं) and the same is endorsed in Vak Suktam, that say's that matter is not capacitated when it is devoid of energy.

Anything can be possible only because it is really the essence of potential energy, Kinetic energy or the different form of energies. Pragmatically the matter of the universe runs by really pretty association of energy (Adi-Shakti).

Shital, we humans and our relationship with the universe is based on the formation and transformation of matter; same is applicable with every object of the cosmos. Ultimately we and this universe are all just manifestations of mass and energy.

We all got broken up back into the natural state of energy to make random universes, particles, etc. Come let's take a move with Shri Brahmaji recitation to Devrishi Narad.

Brahma Ji Said:- Hey Devrishi, tracking the movement of Solar deity once upon in the time-lapse, when the Surya Deva (Solar deity) was annihilated by Lord Shiva, the entire world went in a darkness. Then the whole cosmos was supposed to collapse, the mystic effulgence was appeared in that eon.

Narad Ji asked:- Oh Lord, How interesting, Please let me know the whole epic in detail?

Brahma Ji Said:- Hey Narad listen, once the demons called Mali and Sumali became so powerful by the grace of lord shiv's boon which is the fate of their hard penance.

Thereafter they started to turmoil the world by their deadly acts, all Deities ware worried.

Indra along with other Devas went to Surya Deva and requested to punish the Daityas Mali and Sumali.

Indra said:- Oh Lord of lights, the entire earth is suffering by the highhandedness of Mali and Sumali, they are misusing the boon of lord Mahadev. Both are killing the innocent human, they do this on earth to establish their own kingdom and to rule the world.

Surya Dev Said:- Yes Devraj Indra, I am aware about the doings of Mali and Sumali, you don't worry, I will with my shrill, vanish them and their kingdom.

Brahamaji narrated to Devrishi Narad, thereafter Surya Dev started a calorific intensity by his Heat Polymerization Reaction. The huge and sudden increase in brightness of Sun causes the penetration of radiation in the atmosphere of earth. The innocent being started dying.

Both the demon hailed to lord Shiva. Lord Shiv appeared and said don't panic you are my refugee. Surya Dev with his radiation is even destroying the nature and I don't like Surya dev transformation and denaturalize the system of universe, He is the deity of life giver and by his doings he becomes life taker.

Lord shiva becomes furious and in wrath he shoted his Trident over Suryadev. Suryadev exploded and emitted a solar flare with a huge explosion and the radiation was produced a burst of light and energy and the world went in extreme darkness.

Rishi Kashyap came to Lord Shiva and Said?

Rishi Kashyap Said:- Hey Mahadev, you are the supreme, the Rudra, all know's you are the Lord to make balance in nature. Surya (Sun, the solar deity) is the sourse of life, all the creature may die very soon without solar deity. You being the father of whole world annihilated my son Surya.

It's my curse on you Shiv and the time will come when you have to slaughter (kill) your own son, my curse will come true.

Shiva Said:- Rishi kashyap you are the source of the percept, epistemic, the wise seer. You know no one in this world get rid of their duties (Karma). To keep balance it becomes necessary to vanish the Sun of ego and pride. The Universe will always remain in balance, for every action there is an equal and opposite reaction. Your son falls from his set duties and lost Karma.

Hey Maharishi Kashyap, I accept your curse.

Brahma Ji Said:- Hey Narad, Maharishi Kashyap pacified due to his act of cursed to Lord Shiva, his heart was filled with melancholy, Rishi Kashyap remorse on the act of his ignorance.

Kashyap Said:- Hey Lord of universe, hail you and I am guilty.

Lord Shiv Said:- Hey Maharishi, I will make your son Surya Dev re-irradiated very soon. He will be more responsible after getting alive and he will get abandon from false egos.

* * *

Vijay:- Shital, here Brahmaji was explaining Devrishi Narad.

This was the episode when Surya Deva (Sun) started to eject the bundles of CME's, A **coronal mass ejection** (**CME**) are the significant release of plasma and accompanying magnetic field from the solar corona. The Sun CME's often follow solar flares and are normally present during a solar prominence eruption. The plasma is released into the solar wind, such as groupings of sunspots associated with frequent flares.

Shital, these CME's are so powerful *coronal mass ejection prominence* are the giant bubbles of gas and magnetic fields from the sun, containing up to a billion tons of charged particles that can travel up to several million miles per hour and have been released into the interplanetary medium. This solar material has streamed out through space, and trapped the Earth.

This activity of Surya awakens the wrath of Lord Shiva and Shiv caused Sun to went into **supernova,** *which is an astronomical event that occurs during the earlier stellar evolutionary stages of a massive star's life, whose dramatic destruction was done by lord Shiv by a big explosion. This is the* theory and event of supernova.

* * *

Brahmaji Said:- The universe went to sudden darkness due to devoid of Surya, but before the planets collapse and the gravity breaks. In that darkness the sudden effulgence started to shine the universe. Nobody understood, the source of that faint glow started to illuminate the cosmos.

With the curiosity all Devas get assembled at Swargloka and try to find out who is spreading light and enlighten the universe but do not find anything.They all approached to Shri Narayan along with me and asked Oh Lord. We are unable to detect the glittering source of this celestial and strange light.

Shri Narayan:- Hey Devo, if you want to know this mysterious power, you all go to *the daughter of Rishi Ambhrina,* worship Adi-Shakti and recite the glory of mother **Adi-Parashakti.**

Vijay:- All the celestial beings, the Prajapathi, the Indra of that era started reciting the glory of Adi Shakti.

Devi Adi-Shakti assimilated her sound with daughter of Rishi (Seer) Ambhrna. Her name is Ambhrni; she was the young girl the epitome of chastity and penance. She sounds and Indra along with other Deities. The same Vak (sound) was broken throughout the cosmos.

As soon as he heard, knowledge awakened in him. This was the grace of supreme Vak in the highest heaven. It was Ahi-Parashakti who had spoken through the daughter of Rishi Ambhrna.

Devi's Exclamation:

aham eva svayam idam vada mi,
jushtam devebhir uta manushebhih
yam kamaye tam tam ugram krinomi
tam brahmanam tam rishim sumedham.

It's me who herald and utter these words, all deities and humans will alike admire.

Whomever I love, I make them mighty; Take them to apprehend brahmana, a Rishi and a man of great wisdom.

Devi Suktam or the Vaak Sutam

ऋग्वेदोक्त देवीसूक्तम्
- Rig Veda 10.8.125

ॐ अहमित्यष्टर्चस्य सूक्त स्य वागाम्भृणी ऋषि: सच्चित्सुखात्मक: सर्वगत: परमात्मा देवता, द्वितीयाया ऋचो जगती, शिष्टानां त्रिष्टुप् छन्द:, देवीमाहात्म्य पाठे विनियोग:।

ध्यानम्

ॐ सिंहस्था शशिशेखरा मरकतप्रख्यैश्चतुर्भिर्भुजै: शङ्खं चक्रधनु:
शरांश्च दधती नेत्रैस्त्रिभि: शोभिता।
आमुक्ताङ्गदहारकङ्कणरणत्काञ्चीरणन्नूपुरा दुर्गा दुर्गतिहारिणी
भवतु नो रत्नोल्लसत्कुण्डला॥

देवीसूक्तम्

Vak Uwach

वाक् उवाच

ॐ अहं रुद्रेभिर्वसुभिश्चराम्यहमादित्यैरुत विश्वदेवै:।
अहं मित्रावरुणोभा बिभर्म्यहमिन्द्राग्री अहमश्विनोभा॥ १॥

(Om aham rudrebhi vasubhi scharamyaham adityai rut visvadeveih |
Aham mitra varuno bha vibharmyaham indra Agni ahamasvinobha II)

Devi Exclamation

Om! I move along with the Rudras, Vasus, Adityas and all other Devas.
I bear the Mithra, Varuna, Indra, Agni and the twin Ashwini Devas. [1]

Vijay:- Devi utter aloud, I am the energy which walks with matter (रुद्रेभि)
and moves along with matters (all five states of matter are Rudras, Rudras
of the universe dwells with Energy Ahi Shakti, Rudras in the form of
fundamental forces.) I am with Vasus (वसुभि the master of universal
elements) I am the source of nuclear fission energy of Adityas (आदित्य) the
Sun God (Solar Stars). It's me who abides the Universal-Deities (विश्वदेवै:).

Vijay:- Shital hope you remember, we have already discussed about
Rudras five state of matter (**Solids, Liquids,** Gases, Ionized **Plasma,**
Quark-Gluon **Plasma**) Devi moves along with all these Rudras.

Devi narrated that I am "Kinetic energy" is the energy that is in
motion and causing changes (because Devi sounds I move with). Any
object or particle that is in motion has my kinetic energy based on its
mass (Rudra's) and speed.

Kinetic energy can be converted into other forms of energy, such
as electrical energy and It's her nuclear energy associated with Stars

(Adityas) and performs the phenomena of fission, with chemical energies in Vasu's and thermal energy (Agni).

Devi Said:-

It's me and my potential (potential energy) that bears Mithra, Varuna, Indra, Agni and the two Ashwini (अश्विनी) Devas

अहं सोममाहनसं बिभर्म्यहं त्वष्टारमुत पूषणं भगम्।
अहं दधामि द्रविणं हविष्मते सुप्राव्ये यजमानाय सुन्वते॥ २॥

(Aham Somamahanasam vibharmyaham tvastaramuta pusanam bhagam |
Aham dadhami dravinam havismate supravye yajamanaya sunvate ||2||)

I orbit the Somam who is the destroyer of enemies and the Twashta Prajapathi, Pushan and Bhagam. I give wealth to the performer of the Yajna.

Scientifically, Sacrifice (Yajna) is the ritual, therefore Yagya is the holy offering to pray and please various deities where fire is used as the medium. Fire is one of the elements our body is composed of, so Yagya acts as a link between humans and Deities. In the Yajna who pours the Soma rasa, and who makes the Devas to gain the Havis are due to the sacrifice. [2]

I am the somam as a consciousness that animates and enrich life in plants. Therefore I am the "Universal Life Energy" as a Soma.

Scientifically, Devi stated that it's me who strengthen Twastha by which he accomplish the **process of Photosynthesis,** *by which plants, some bacteria and some protistans use my energy from sunlight to produce glucose from carbon dioxide and water.*

I am the energy in the form of ignition (दधामि **Ignition energy (IE)** is the amount of energy required to ignite a combustible vapour, gas or dust cloud) one who gives elixir to Devas and Devas receives **Havis** (food of gods) by my energy.

Havis is the sacrificing phenomena of food (द्रविणं) and transformation of food energy by the act of fire energy to the cordial oblation for gods.

अहं राष्ट्री संगमनी वसूनां चिकितुषी प्रथमा यज्ञियानाम्।
तां मा देवा व्यदधुः पुरुत्रा भूरिस्थात्रां भूर्यावेशयन्तीम्॥ ३ ॥

(Aham rastri sangamani vasunam chikitusi prathama yagyiyanam| Tam
ma deva vyadadhuh purutra bhuristhatram bhuryavesayantim ||3||)

Devi further said:- Hey Bahushrut Indra,

I am the Queen of the Universe as primordial energy; I give wealth
to those who worship me. I am the all-knowing one and the prime
one among the worshiped deities. I enter in everybody as 'Atma' by
the processes of transformation, taking various forms and I diversify
myself into various manifestations, in various ways. Hence, the myriad
Deva's are incorporated within me.[3]

मया सो अन्नमत्ति यो विपश्यति यः प्राणिति य ई शृणोत्युक्त म्।
अमन्तवो मां त उप क्षियन्ति श्रुधि श्रुत श्रद्धिवं ते वदामि॥ ४ ॥

(Maya so annamatti yo vipasyati yah praniti ya iim srnotyuktam |
Amantavo mam ta upaksiyanti srdhi srta sradhivam te vadami ||4||)

That one who eats food, who sees, breathes, and hears whatever is
said, they does all only through me (my energies). Those who do not
understand me, die. O dear one! (To the worshipper or devotee), give
your attention to what I sound with concentration. [4]

अहमेव स्वयमिदं वदामि जुष्टं देवेभिरुत मानुषेभिः।
यं कामये तं तमुग्रं कृणोमि तं ब्रह्माणं तमृषिं तं सुमेधाम्॥ ५ ॥

(Ahameva svayamidam vadami justam devebhirta manusevih | yam kamaye
tam tamugram krnomi tam brahmanam tamrsim tam sumedham ||5||)

"All these are me (and various manifestations of mine). I am the one
worshipped by Devas and the earthly beings. If I like someone (for his
meditation towards me), I make them the greatest alike Brahma, the
most intelligent as a Sage, and as a Self-Realised soul. [5]

अहं रुद्राय धनुरा तनोमि ब्रह्मद्विषे शरवे हन्तवा उ।
अहं जनाय समदं कृणोम्यहंद्यावापृथिवीआविवेश॥ ६ ॥

(Aham rudraya dhanuratanomi brahmadvise sarave hantava u | Aham janaya samadam krnomyaham dyavaprthivi aa vivesa ||6||)

I bend the bow of the Rudra (I am the form of elastic energy, bend and stretching the bow) to kill all those enemies (anti matter) who detest all good things. I fight these bad (Antimatter) elements/enemies only for the people to adorn the creation of Brahma. *I enter, pervade and persist throughout the earth and the Space or Sky with my various aspects.*[6]

अहं सुवे पितरमस्य मूर्धन्मम योनिरप्स्वन्त: समुद्रे ।
ततो वि तिष्ठे भुवनानु विश्वोतामूं द्यां वर्ष्मणोप स्पशमि॥७॥

(Aham suve pitaramasya murdhan mama yonirapsvantah samudre |
Tato vitisthe bhuvananu visvotamum dyam varsmanopasprsami ||7||)

I created the sky (inflation of Space as a Vishnu was done by the Vast power of my Adi-Shakti form), which is (as a shelter) above the earth and which (Lord Vishnu) is fatherly for all beings. My creativity (power) is within the Ocean. By that, I am present in all the worlds (different form of energies). And I touch the sky with my roomy body. [7]

अहमेव वात इव प्रवाम्यारभमाणा भुवनानि विश्वा।
परो दिवा पर एना पृथिव्यैतावती महिना संबभूव॥८॥

(Ahameva vata iva pravamyarabhamana bhuvanani visva |
Paro diva par ena prthivyeitavati mahina sam vabhuva ||8||)

When I start creating all the worlds (भुवनानि), I function like the air (It should be noted here: at the beginning of the universe there is only Hydrogen and Helium). The gas energy during the epoch of (Anhad Naad), 13.75 billion years ago – The Anhad Naad (Big Bang) – (It is not known what triggered the Big Bang for modern cosmologists). But Vedas says and we believe, the process called inflation is the origin of **Shri Vishnu** as a space happened in the fraction of a second by the cause of Naad by the act of vibration between Naad and Bindu.

That's my strength (as energy) which holds the entire living creature together.

Vijay:- There was a strange type of **vacuum energy,** this is what cosmologist of our era believes that caused the universe – the volume of space itself – to expand by a factor of 10^{78} in a fraction of a second.

It's not a strange energy virtually that's **Devi Adipara-Shakti;** she is the cause for the inflation of space.

But we know as the Devi Sounds in Vak Suktam that I am Adi-Parashakti who caused the inflation (Begat) and created the universe and expanded. I am taller and higher than the Sky. I am greater than this earth because all elements like earth created by my power. Such is my valour, strength, prowess and greatness.

Note: Vak - the daughter of Rishi Ambhrina, she had connection with Devi. It's her highness and her **Sound energy** *is a form of* **energy** *that is* associated with vibrations of matter (Rudra). It is a type of mechanical wave which means it requires an object to travel through. This object includes air and water. **Sound** originates from the vibrations that result after an object applies a force to another object.

Vijay:- It's our Vedas, our ancient culture and the civilization which shows us that everything is made up of energy. It's the building block of all matter. The same energy that composes our body is the same one that composes the bricks of the house we live in, our accessories, our phone, these animals, trees, birds, celestial bodies both known and unknown and so forth. It's all the same energy in different forms. She constantly flows in the form of various energies from eternity.

Adi-Shakti in different form of energy is everywhere, and what modern science has shown us is that energy is neither created nor destroyed was already described long back in Vedic time.

Everything in this universe is made up of the same stuff. It is just present in different forms and shapes by the mean of Mass, Energy, Matter and Space.

Modern science in spite of day to day research cannot solve the mystery of nature (Prakriti). We have to solely believe in the power of divinity Adi-Shakti and Shiv. We should have strong trust in Shri Narayana.

Shital:- Wow, It's great to know the secret of Vak Suktam, her Excellency Devi Para-Sakti is narrating the classification of her own world of sovereignty therefore she rules the eternal cosmos.

Vijay:- The state of dark energy era remains for a long time, but all "Devas" observed a faint glow of Adipara-Shakti during this epoch.

After this then Shiva started to recall all his aspects to alive the Sun God. From Shiva appeared his five forms Ishana, Tatpurusha, Aghora, Vamadeva and Sadyojatha.

Ishana by his power of gravity and his all son's contributed by mean of their fundamental forces.

The Jati the force of gravity applied by Ishan Mahadev acted upon the remnants of Sun and because of his power of gravity the asteroids and comets club together again.

Inshan Shivas 2nd son Mundi generated electromagnetic forces once again and Mundi by its power of electromagnetism the force responsible for the way by which matter generates and responds to electricity and magnetism.

While Ardhmundi and Shikhandi developed strong and weak force, Strong Nuclear Force binds the nucleus of an atom together. The Shikandi's Weak Nuclear Force is responsible for certain kinds of radioactive decay.

Then the Rudra Sadyojata, aspect of Mahadev Shiv and his descendant generation Sunand, Nandan, Vishwananda and Upnandan started saturation of gaseous Hydrogen, Helium, Deuterium and Hydrogen bonding.

Sunand given its gas 91 % of gas was hydrogen which was converted into energy in the sun's core. The energy moves outward through the interior layers, into the sun's atmosphere, and is released into the solar system as heat and light.

In the sun's core, gravitational forces of Jati create tremendous pressure and temperatures. The temperature of the sun in this layer is about 27 million degrees Fahrenheit (15 million degrees Celsius). Sunand compressed and fuse together his own Hydrogen atoms, created helium which is the aspect of Nandan. This process is called nuclear fusion. As the gases heat up, atoms break apart into charged particles, which turn the gas into plasma.

This is how the Sun comes once again into existence by the grace of Lord Mahadeva Shiv.

The Vishwanandan and Upnandan help once again to associate the Sun in penal of its Solar system.

Jati the Gravity once again made the planets to orbit the Sun. This way the solar system functions and got their super soul with the pious Sun.

Shital:- That's right, exactly modern Science also explains the same, the only difference is now we know the cause behind every anecdote through Vedic science.

Vijay:- People shoud not think this as my imaginary cult, it's the testimonial theory which is supported by New work from Carnegie's Alan Boss who offers fresh evidence supporting this theory, modeling the Solar System's formation beyond the initial cloud collapse and into the intermediate stages of star formation. It was published by the Astrophysical Journal. This Journal states our Solar System's formation was triggered by a shock wave from an exploding supernova. The shock wave injected material from the exploding star into a neighbouring cloud of dust and gas, causing it to collapse in on itself and form the Sun and its surrounding planets.

This is how the Sun our solar deity again brought in existence after supernova caused by Lord Mahadeva.

Shital:- This is something very great, now I want to understand how our earth was formed because the current theory of science say's the earth was formed from the solar nebula. Can you please tell me from the record of Vedic or Puranic science?

Vijay:- Exactly, Earth origin is the sequential in continuation of Sun's origin. As you know Lord Shiva Shankar made Sun (Surya Deva) as space is filled with gas and dust. Most of the material was hydrogen (Sunand) and helium (Nandan), but some of it was made up of leftover remnants from the violent deaths of star our Surya Deva caused by Lord Shiva. Waves of energy Adi-Shakti roam and ramble through space pressed clouds of such particles closer together, and gravity of Jati causes them to collapse in on each other. As the material drew together, gravity of Rudra **Jati** once again caused it to spin. In the centre, the material clumped together to form a protostar that would eventually became our sun.

Shri Brahma Said:- Hey Devrishi Narad, Now you listen how the earth came in existence.

Narad Ji Said:- Oh how fortunate I am, eagerly waiting to listen the story of mother earth origin?

Brahma Ji Said:- When Lord Shiva accumulated the remnant of earlier dead Star (Surya) and recall Sunand and Nandan by the help of Jati (Gravity) the nuclear fission started in the core of sun, this way Surye deva appeared once again.

There was debris of heavy elements scattered all around the solar protoplanetary disk. When Rishi Kashyap and Sapta Rishis (Seven Sage's) knows the significance of Shivleela, they invoked to Lord Shiva.

Maharishi Kashyap said:- Oh lord Salutation to you, you are the ocean of mercy but this stuff remnant are harming the universe. There is no one on whom 'Surya Dev' will grace his rays of light, what is the use of these leftovers relics of early dead star.

After listening to them, a faint smile passed on Lord Shiva's lips and he said.

Lord Shiva Said:- Hey Sept Rishiyon, the remnant leftover seems to you is of no use now, but in future that will be the prime source of life, living and habitat.

After saying this, Lord Shiva from his strong Gravity (Jati) pulled the scattered debris leftover with frequent volcanic activity towards his own physique merged and the outcome of his activity created earth. The temperature was very hot at that time.

The tremendous energy needed in the processes which was applied by the wisdom of Shree **Bhuvaneshwari** Devi. Since lord Shiva is Shambhu, created earth in association of **Bhuvaneshwari**. Hence earth then got the name Bhu Devi, the name Bhu Devi is derived from Shambhu and **Bhuvaneshwari**.

Lord Shiva **Jati** as a gravity exist at the core of Earth (Bhu Devi) hence having strong gravity started attracting the bodies towards her and form a vast structure.

Shital:- There is one doubt perceived now, As you told that due to the effects of solar storms in the forms of CME's of earlier sun, the earth suffered from a big loss. Are that earth and this earth the same because you told the earth was created by Shiva with the remnant of Surya (Sun) after he went to the supernova?

Vijay:- There are numerous Planets, Suns and Stars in the universe. That earth which got suffered by the coronal mass ejection's (CME's) of Sun is different. That might be Gliese 667Cc or any one from the series Kepler or may be any other.

Shital:- You means that was a different planet like our earth, If yes I want to know more about Gliese 667Cc, it's habitable zone and whether life on Gliese 667Cc was completely vanished or still the inhabitants lives on Gliese 667Cc?

Vijay:- Yes modern science say's it an **Earth analogue** also referred as **Earth-like planet**, with environmental conditions similar to those found on the planet Earth.

There may be variation in environmental and the climatic condition of earth analogue planets based on their atmosphere.

I will explain you this with the reference of journal published by Physics.org on JUNE 25, 2013

Three planets in habitable zone of nearby star

Gliese 667C, an exoplanet with the habitable zone

This artist's impression shows the view from the exoplanet Gliese 667Cd looking towards the planet's parent star (Gliese 667C). In the background to the right the more distant stars in this triple system (Gliese 667A and Gliese 667B) are visible and to the left in the sky one of the other planets, the newly discovered Gliese 667Ce, can be seen as a crescent.

A team of astronomers has combined new observations of Gliese 667C with existing data to reveal a system with at least six planets. A record-breaking three of these planets are super-Earths lying in the zone around the star where liquid water could exist, making them possible candidates for the presence of life. This is the first system found with a fully packed habitable zone.

Gliese 667C is a very well-studied star. Just over one third of the mass of the Sun, it is part of a triple star system known as Gliese 667 (also referred to as GJ 667), 22 light-years away in the constellation of Scorpius (The Scorpion). This is quite close to us—within the Sun's neighbourhood—and much closer than the star systems investigated using telescopes such as the planet-hunting Kepler space telescope.

Previous studies of Gliese 667C had found that the star hosts three planets with one of them in the habitable zone. Now, a team of astronomers led by Guillem Anglada-Escudé of the University of Göttingen, Germany and Mikko Tuomi of the University of Hertfordshire, UK, has re-examined the system. They have added new HARPS observations, along with data from ESO's Very Large Telescope, the W.M. Keck Observatory and the Magellan Telescopes, to the already existing picture. The team has found evidence for up to seven planets around the star.

These planets orbit the third fainter star of a triple star system. Viewed from one of these newly found planets the two other suns would look like a pair of very bright stars visible in the daytime and at night they would provide as much illumination as the full Moon. The new planets completely fill up the habitable zone of Gliese 667C, as there are no more stable orbits in which a planet could exist at the right distance to it.

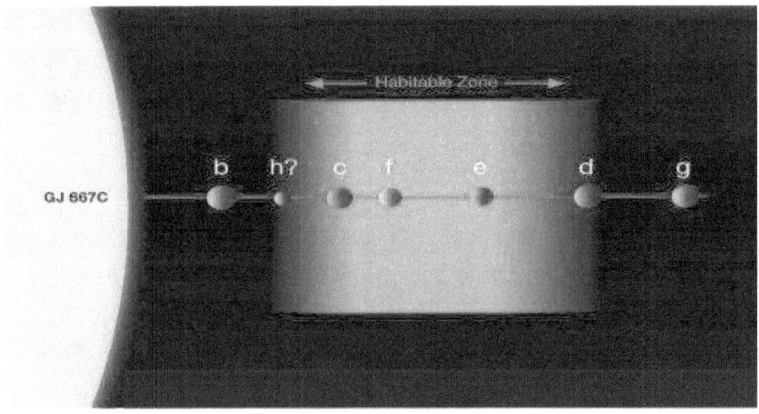

**Planetary System around the star Gliese 667C in which
three planets c, f and e are Super-Earths**

This diagram shows the system of planets around the star Gliese 667C. A record-breaking three planets in this system are super-Earths lying in the zone around the star where liquid water could exist, making them possible candidates for the presence of life. This is the first system found with a fully packed habitable zone. The relative approximate sizes of the planets and the parent star are shown to scale, but not their relative separations.

"We knew that the star had three planets from previous studies, so we wanted to see whether there were any more," says Tuomi. "By adding some new observations and revisiting existing data we were able to confirm these three and confidently reveal several more. Finding three low-mass planets in the star's habitable zone is very exciting!"

Shital:- That's great, these are amazing facts, it means there might be still life on **Gliese 667Cc.**

Vijay:- Certainly, but this rocky planet was suffered the huge attacks of CME's and have big loss of life which evoke the wrath of lord Shiva.

Because the earth is not just a planet, it's among one element from basic five elements (Panch Tatva) which was created at the early phases of the universe by Lord Shiva in the form of Tatva Lingum. Thus many planets in the universe are made up of elements called Prathvi Tatva. This we already discussed in our previous discussion.

Shital:- But Prathvi is our planet and as you told Prathvi is an element, What's the relation?

Vijay:- Since Prathivi was created as one of the basic five element in the beginning of Sarga by Brahma Ji. Our planet **Bhumi Devi** is made up of **Prathvi Tatva** (dust particles), it is also known as Prathvi Devi. The earth doesn't have any logical resemblance it's the name of modern language.

This Earth which was created by the remnant of Surya (supernova star), is our mother earth called **"Bhumi-Devi".** The growth of Bhu-

Devi (earth) involves many stages of evolution which gave rise to an atmosphere which is suitable to propagate life and finally this planet got the habitable zone.

Bhumi Devi is also known as **Kashyapi** because she was created on the request of Maharishi Kashyapa hence she is known to be the daughter of Kashyapa and then nourished by cosmic matrix of Aditi, so Aditi is the mother of Bhu-Devi.

Brahma ji said:- Hey Narad, That's way Sruthi Says:

सत सृष्टि तांडब रचयिता नटराज राज नमो नमः

This means that it was Lord Shiva's Cosmic Dance (Srusthi Tandav) by which he creates and epithet the cosmos (Rachaita means the Creator, Lord Shiva is creator in a Natraj form).

Briefly when Lord Shiva was in anger he has done the Tandav dance to razed solar deity and then he was praised by Maharishi Kashyap. On this Shiva performed Natraj dance to create once again the solar deity as Surye Deva and the earth too.

Shital:- So this processes is applicable on the origin of all the stars and planets.

Vijay:- There are many reason for the birth and death of stars and planets. The universe is filled with countless celestial objects. Variable processes are involved in the origin of Stars and Planets.

Shital:- Can you elaborate some more facts on the birth of Stars and Planets?

Vijay:- Lets go back in the Saptrishi Mandal or the group of seven star, Now the universe is so vast because of continuation of inflation of space. We already knows that Shri Vishnu as a space is expanding continuously from big bang to now.

When we talk of Space during the very first generation Stars- Population III Stars, the Space expansion during the the phase of first-generation stars was slow in comparatively today. These 1st generation stars were likely very large, about a hundred times the mass of our Sun but in small population. Because of their size, they had short lives which ended as supernovae and heavy populated universe with linage of second generation and same way in the 3rd generation of today's universe.

Let's talk about the 1st Manasputra of Brahmaji, He created this star from his eyes, this happened when Shri Brahma Ji involved deep in the processes of creating the universe. A blazed ray of lights emerges from his eye, which was the sparked light of Maharishi Marichi. Immediately after the origin he asked his father Brahmaji.

Marichi Said:- Hey pita, What is my role in this creation?

Shri Brahma ji:- Hey Putra (son) Marichi, you contribute in the processes of creation. You will be efficient to give birth to the son, who will be the great sage and whose progeny will fill this universe in the form of various creatures.

Vijay:- As said by Lord Brahma Ji, Kashyapa becomes the Son of Marichi. Therefore the Sage Kashyap is none other than a constituent manifestation of Lord Rudra and Maharishi Kashyap is known as the 'Lord of Creatures' because Prajapati Daksha married his thirteen daughters (Aditi, Diti, Kadru, Dhanu, Arishta, Surasa, Surabhi, Vinita, Tamra, Krodhavasa, Ida, Khasa and Muni) with Maharishi Kashyapa and He was the father of the Devas, Asuras, Nagas, humans and other creatures from these wifes.

Same way the celestial personalities are begotten from other Seven Sage's Like Maharishi Angira made Jupiter (Brahspathi-The planets of wisdom and the teacher of all heavenly Deities. Same way Planet Venus, the Shukra got origin from Maharishi Bhrigu who later rewarded as the mentor of Daityas.

Same is the case of planets and galaxies which has multiplied in many bodies.

5.18 Adityas, beyond Our Solar System

Shital:- We know, our solar system is illuminated by Sun God Surya, As modern science and Vedas both says there are numerous Galaxies with uncountable solar system's and several stars, planets and other celestial objects. I want to know about other exoplanets and their source of light. What about the life on those planets.

Vijay:- Our sun is one of at least 100 billion stars in the Milky Way (आकाशगंगा), a spiral galaxy about 100,000 light years across. Same way numerous stars are found in other galaxies.

The Sun is one of the Aditya which we can see; apparently there are more Adityas in other planetary system. Adityas are those who begotten from the cosmic matrix, the Adi-Shakti in the form of Infinite energy (Aditi) the association of Maharishi Kashyap.

The Milky Way is part of the Local Group, a neighbourhood of about 10 million light years across, consisting of more than 30 galaxies that are gravitationally bound to each other. Aside from our galaxy, the most massive one in this group is Andromeda, which we can see even with our naked eyes, which appears to be on course to collide with the Milky Way in about 4 billion years.

Shital:- You mean all the shining objects like stars are the Aditya's, isn't it?

Vijay:- No, Aditya's are those who got originated by the Aditi and Kashyap, while many stars and planets etc. are the succession of other supernove or by any other mean.

Remember Adityas are the ingredients of the Cosmic Matrix Aditi. All these personalities are the opulent expansions of the Supreme Personality of Godhead, Shri Narayan, in the form of the solar deities.

As Aditi is the infinite cosmic matrix, Adityas represents to perform many tasks and responsibilities of the universe.

Rigveda, describe Adityas are seven celestial deities, sons of Aditi. they are: Varuna, Mitra, Aryaman, Bhaga, Ansa (Amsa), Dhata, Indra.

In addition to this Purana added them to be twelve as listed. Parjanya (Savitr), Tvashtha, Vishnu, Pushya, Vivasvan.

The Scientists all around the world observed common node of discoveries that Stars are the source of light, energy with the combustion of helium (Nandan). Knowledge about stars are more advanced in Vedic era, thoughts are not limited only on the concept hydrogen in core which is converting into helium by nuclear fusion.

Adityas are expedient to perform various tasks on their associated planetary system. Like on our earth (Bhu Devi), Surya is the source of life in different forms.

Because of him, plants and other organisms convert light energy into chemical energy that to fuel the organisms activities (energy transformation). So this is the nature of Aditya called Tvashtha (life of plants) which convert the light energy of Sun into chemical energy as a food for plants.

Shital:- So you mean Tvashtha is not a different or distinct shining entity other than our Sun.

Vijay:- Tvashtha is one among other Adityas and definitely a distinct shining Star but he helps in the processes of photosynthesis on various planets by the Association of Adi-Shakti as we have seen in the commentary of Vak suktam by Devi.

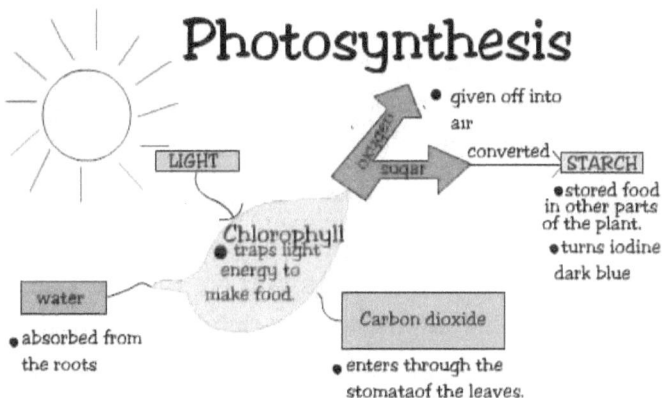

Aditya Tvashtha performs photosynthesis in the vegetative organism

All living organisms require food, in plant food processing is not an automatic process, plant can make their own food, this is called the mechanism of photosynthesis which is performed by the son of Aditi called Tvashtha and he have to do this tasks at infinity. Aditi is infinite energy and his son's Adityas are multiple task performer. It is a complicated process, but basically, carbon dioxide and water are converted to glucose (a simple sugar) and oxygen.

Photosystem found in plants and cyanobacteria, uses photons (Savitr) of light to break apart molecules of water extracting electrons to fuel the photosynthetic conversion of light and water into chemical energy for cellular functions.

Aditya Twasta not only helps in creating food for plants, also creates various birds and animals.

5.19 Binary System of Double Star Mitra & Varuna

Mitra and Varuna are the twin brothers and the Adityas in the linage son of Kashyapa and Aditi.

Mitra-Varuna are the dual divinity because it expresses the complementarities of the pure interdictory always have mutual alliance and the impure transgressive Varunic poles of Vedic sacrality, also translated into the opposition between the upper and nether worlds of a dualistic cosmos.

Rigveda refers, Mitra and **Varuna** as two twin deities and both considered as Adityas, or deities connected with the Star; and they are protectors of the righteous order.

Shital, you should noted and this phenomena has to be deemed comprehend, It's a wrongly perception that all Adityas are Sun god, it will be the genuine statement that Adityas are the Star's and each star in the group of Aditya's composition are rich in different material. Sun is rich in helium and hydrogen, Mitra and Varuna is rich in composed oxygen that means water.

Binary System of Twin Star Mitra & Varuna as VFTS 352

Their association is so close and predominantly linked that only one single hymn of RV 3, 59 is addressed to Mitra separately.

About 160 000 light-years from Earth (Bhu Devi) in the Large Magellanic Cloud lies a double star system Mitra-Varuna. Modern science named it as VFTS 352 is the hottest and most massive double star system till date, with components so close that they touch each other.

In fact, the Mitra-Varuna stars are so close that their surfaces overlap and a bridge has formed between them. Mitra-Varuna (VFTS 352) is not only the most massive known stars in this tiny class of 'overcontact binaries', it has a combined mass of about 57 times that of the sun, but it also contains the hottest components — with surface temperatures above 40,000 degrees Celsius.

Modern Astronomers found that, extreme stars like the two components of VFTS 352, play a key role in the evolution of galaxies and are thought to be the main producers of elements such as oxygen. The same hymn found in RV that Mitra supports heaven and earth that means the same, evolution of galaxies (Galaxies are the clusters of Stars Planets and other celestial's objects).

In the case of VFTS 352, however, both stars in the system are of almost identical in size.

Varuna in the Veda is not an Indian Neptune, neither is he precisely, as the European scholars at first imagined, the Greek Ouranos, the Sky. He is the master of an ethereal wideness, an upper ocean, of the vastness of being, of its purity; in that vastness, it is elsewhere said, he has made paths in the pathless infinite along which Surya, the Sun, the Lord of Truth and the Light can move.

Material is, therefore, not sucked from one to another, but instead may be shared. The component stars of VFTS 352 are estimated to be sharing about 30% of their material.

In Vedas there are different group of star in each 'manvantar'. Manvantar is the Vedic astronomical time within each Kalpa (aeon), a "day of Brahma" and the present *Śveta Vārāha Kalpa* in which we live today, where again 14 Manvantaras add up to create one Kalpa.

Each Manvantar has different set of seven Stars like we have today in Ursa Major. The life of those Stars ended with supernovas and produced the numerous progeny in the form of new stars, planets, galaxies, other celestial bodies and even the **Saptrishis** in human form.

The below is the scientific spiritual and holy prayer (Stuti) in which the two cosmic dance of Shiva symbolize the dissolution, Sustainability and creation of universe.

The two aspect of dance by Shiv are:

1. Tandav – the annihilation of universe

2. Natraj – the ecstatic dance is the aspect of recreating the cosmos.

सत सृष्टि तांडव रचयिता नटराज राज नमो नमः।
हे आद्य गुरु शंकर पिता नटराज राज नमो नमः।

Sat srushti tandav rachayita, Natraj raj namo namah:
Hey Adhya guru Shankar pita, Natraj raj namo namah!

गंभीर नाद मृदंगना धबके उरे ब्रह्मांडना
नित होत नाद प्रचंडना नटराज राज नमो नमः।

Gambhir naad, mrudangna, Dhav ke ure brahmandna:
Nit hot naad prachandna, Natraj raj namo namah!

सिर ज्ञान गंगा चंद्रमा चिद ब्रह्म ज्योति ललाट मां
विष नाग माला कंठ मा नटराज राज नमो नमः।

Sir Gyanganga chandrama, chid bramhjyoti lalat ma:
Vish Naag mala kanth ma, Natraj raj namo namah!

तवशक्ति वामे स्थिता हे चन्द्रिका अपराजिता ।
चहु वेद गाएं संहिता नटराज राज नमो नमः।

Tav Shakti vaame sthita, Hey chandrika aprajita:
Chahu Ved gaye Sanhita, Natraj raj namo namah!

With this Stuthi we will now proceed to the further growth of the universe, see as of now numerous universes come into existence by the grace of Goddess **Bhuvaneshwari,** all the universes are hold by **Garbhodakshayi Vishnu Shri Narayan.**

Shital you remember in our early discussion I have narrated you, how Prajapathi Brahma has created the living beings and other Tatvas. But in the process, his creation was not getting alive after all his efforts. So once again we will talk about the same episode of that tale.

In Vedas, as well many scientists stated that life came from a distant world or might have fallen with comet or asteroid on earth. Panspermia is the hypothesis that life exists throughout the Universe, distributed by space dust, meteoroids, asteroids, comets, planetoids, and also by spacecraft carrying unintended contamination by microorganisms. Distribution of life may have occurred spanning galaxies, and so may not be restricted to the limited scale of solar systems.

Reference Says, Some organisms may travel dormant for an extended amount of time before colliding randomly with other planets or intermingling with protoplanetary disks. Under certain ideal impact circumstances (into a body of water, for example), and ideal conditions on a new planet's surfaces, it is possible that the surviving organisms could become active and begin to colonize their new environment.

Reference:- https://science.nasa.gov/solar-system/big-questions/ how-did-life-begin-and-evolve-earth-and-has-it-evolved-elsewhere-solar-system

So I will narrate you the same which was once asked by Seer Narad Ji, to Prajapathi Brahma Ji.

Devrishi Narad Ji asked:- Hey Pita, How this universe is perceiving, sustaining and being annihilated. What are the reasons and how life begins I want to cognize and understand the spectator Theory of Knowledge?

Prajapati Brahma Ji Said:- Hey Narad, You are always roaming in the search of knowledge and cognizing that to those who are engaged in the subject of welfare.

We were instructed by Mahadev Shiva in an aforesaid eon, therefore I was ecstatically, thereafter taken permission from Shri Hari Vishnu decided to penance to start the creation.

Shri Hari Vishnu also evanesces from thereafter addressing me. He physically went away out of this universe and stay in Baikuntha (बैकुंठ) that is his permanent abode.

By reminiscence of Sada Shiva and Shri Vishnu Ji, I started the processes of creation. I with the help of auspicious water and other Panch Tatvas (the five elements air, water, fire, earth and space) created the universe of twenty four elements.

But the vast Oosphere the egg (und) was inert without life. By seeing no life I was very suspicious and started hard penance. After a long penance of twelve years of that time scale, Shri Hari Vishnu was pleased and appeared in front of me.

Shri Vishnu Said:- Brahm Dev open your eyes.

The mellow sound opened my eyes and I found Narayan in front of me. He further said,

Shri Vishnu Said:- Hey Brahman I am very much pleased by your hard penance, You can asked your desirous boon. I am capable to accomplish your desire.

Brahma (I said):- Oh Great Lord, of the cosmos, O God of gods, O Protector of those who seek your shelter, salutation to you, Hey Narayan I have created this Cosmic Und of twenty four elements but in spite of my determination and hardwork I can't make it alive?

Hey, Parmeshwar, You are the Almighty Lord, outshining your glory, pervaded everywhere; ample is your love for cosmos constitution.

Oh Lord, you begin where comparison ends, you are limitless, boundless only you can reconcile this universe, please make alive this group of twenty four elements created by me.

On my request Shri Hari Vishnu appeared as endless and infinite form, at that time the Param pursha Shri Narayan posses billions of Heads, Hands, Legs and Limbs and he then surrounded that cosmic egg (Und) from all sides.

I recited a holly prayer to lord Shri Vishnu after which he entered with high volt strike thunderbolt light energy into the cosmic egg.

Due to the effect of that lightning phenomena of Shri Narayan the cosmic egg made up of twenty four elements become animate, conscious and alive.

> The **lightning origin of life theory** postulates that billions of years ago a lightning bolt (or intense period of electrical activity) hit a warm pond or primordial soup and thus triggered, sparked, or activated the first reactions of life, or something along these lines.

Now there is fully grown up atmosphere from heaven to Patal loka was sustained by Shri Narayan Vishnu that's the reason Shri Vishnu is also called Viraj-Pursha.

This way the Inanimate, Inert, Insensate and Radix planet earth get habitable zone.

The Kailash Mountain is the abode of Shiv-Shakti and Baikunta Kseer Sagar (Ocean) is the abode of Shri Narayana are beyond any perception.

Shital:- I remember in our early discussion your pondering with this ultimate truth of life and the same resemblance theory was given by modern science in the aspect of origin of life.

Vijay:- Few years back in Bhopal a young student named Mahima has asked my friend Devendra Pankaj (Dev) about the theory of origin of life. I was standing beside my friend Dev. He explained nicely the lightning origin of life theory of naturalist Charles Darwin to that young girl. Dev is representing **the** best in **the** profession and set **the** highest standard for best practice as a teacher.

On that day I thought that some time when I will get opportunity I will put forth this modern theory which was explained in our Vedas and Puranas, fantastically at very bygone (During Vedic Period) with all the evidential record.

व्याप्त कर लिया । मेरे द्वारा भलीभाँति स्तुति
की जानेपर जब श्रीविष्णुने उस अण्डमें
प्रवेश किया, तब वह चौबीस तत्त्वोंका
विकाररूप अण्ड सचेतन हो गया । पातालसे

We will see what naturalist Charles Darvin says about the theory. *"The original spark of life may have begun in a warm little pond, with all sorts of ammonia and phosphoric salts, lights, heat, electricity, etc. present, so that a protein compound was chemically formed ready to undergo still more complex changes"*.

In the continuation of Vedic discussion of Prajapathi Brahma Ji and Dev Rishi Narad we will see how scientifically Prajapati Brahma brought up living creature in this universe is his science, not a magic.

Brahma Ji Said:- Hey Devrishi, Then I have decided by the wish of Lord Shiv to organize a creation while staying in Satya Loka.

Hey Son, in the processes of creation due to ignorance, I have created a creation of Tamo-Guni which is called **Avidhya-Sarga** (अविद्या)- (Ignorant Creationist- I certainly hope that my ignorance is the bliss of lord Shiva and that I truly believe).

There after I was happy and delightedly engaged in meditation of creation by the bliss of lord Shiv Shambhu, Hence I have created Vegetative Existence ('स्थावर') like trees and plants which were called as a main Sarga (and is the first Creation). Thereafter I realized that, trees and plants are unable to do their self welfare.

Then I have created Second Sarga which is called **Triyak-Sarga** (त्रियक) are bacteria's (जीवाणु), virus (विषाणु) and insects (कीट) this Sarga was also unable to do any self welfare.

Those all Sarga's are effortless in the aspect of their own wellbeing I have created the 3rd Sarga known as **Urdhavstrota** Sarga or the Upper Sarga and that is superior the celestial Sarga famous as **Dev Sarga.** This Sarga is creation of Divine celestial Deities which is the planet of glory (Swarga).

I thought those to be also an effortless in the welfare of the cosmos, then I again started meditation and by the bliss of lord Shiva created the Sarga called **Awarktastrota** Sarga having the qualities **Rajas Guna** and that are Human Being and really engaged in the activities of Humanity and welfare of the cosmos.

Then I have created Sarga of Arial spirits. My creation involves five types of Sargas.

Vijay:- This is not so simple that the Brahma Ji created Sarga by seated on his abode of Satya-Loka, he is the master of creation, Master of science. The world which we visualize is his research, design and the creation of mastermind Brahma Ji.

Shital:- No answers today, only a perennial question. But it's one of the most interesting and meaningful questions I can possibly ask.

Where does life come from, for which you answered Brahma created life by different mean of **Sarga** (Creation)?

How did life reached on Earth? There was no life on the planet we live today in the beginning, to the vast abundance of life in different mean we see now, it will be the endeavor to understand life's beginnings?

Vijay:- That's astonishing, your curiosity overwhelmed me as it is somewhat the same, which was one of the best and greatest story of BBC Earth's 2016. So listen, this is the story of our quest to discover our ultimate origin.

Charles Darwin first published modern theories of evolution – that all life on Earth is related; adapting and changing over time. Look at any two creatures on Earth and you can trace them back to a common ancestor. Humans and chimpanzees share a common ancestor from at least 7 million years ago.

The Vedic Science has the more shreds of evidence which say, that every existence in the cosmos have primordial common ancestors. As everything was created by Brahma Ji and Shri Brahma Ji sprouted from Narayana and Shri Vishnu Narayan begat from Shiv and Shiva. So Shiv and Shiva (Shakti) is our common ancestor. That was the beginning of celestial life.

So the answer to the question, from where did we get life on Earth, is that life began at different galaxy were Brahma seated and delivered on to earth. Vedas have figured out how the life has sprung up first.

Let's come to the Vedic story before coming to the discussion of various modes of creation (Sarga). Let me narrate you the story of Shiv and Rudras before the beginning of the above discussed *Sarga* (vegetative life, micro-organic or any immortal life) Braham Ji was suffered from the sorrow of his own creativity as a resulted four kumars **Sanak, Sanandan, Sanat and Kumar** have taken birth from his heart *desire* (Mann) and they all are brilliant, knowledgeable and talented. But as discussed earlier they all refused Brahma Ji to help in the processes of creation and get engaged in divine salvation (nirvana).

Brahma Ji Said:- **Hey Narad,** This anecdotal story I have narrated you in our earlier discussion, I will describe you again and disclose more secret of my creation.

When I have not found consciousness in the universe created by me, then I started meditating Shri Hari Vishnu and he appeared, by guided me he told.

Shri Narayan Said:- Hey Prajapathi Brahman, don't worry, now you should penance to please Lord Shiva.

I (Brahma) Said, I will do as you suggested and I started hard penance. Lord Shiva shortly got pleased and graced me by making an appearance in Shiva-Shakti Ardhnarishwar Swaroopa (Sanskrit: अर्धनारीश्वर is a composite androgynous) form. I recited hymns to praise Lord Shiva.

The Ardhnarishwar swaroopa is the epitome, symbolizes the ultimate truth and unity of Shiv and Shakti. Ardhnarishwar is the physical appearance of singularity and the association of Pursha and Prakriti.

I recited a Sruthi in the glory of Shiva and requested him to create the different types of beings.

The glory of Ardhnarishwara

Brahma Ji Recited:-

Salutation to you o Lord, Great Lord, O God,

You are the lord of lords O Shankar, acclaim you.

O Devi you graced this nature, Yor are the habitude principle of nature,

You reside even free from nature, Universal beauty you are.

O Devi you have a great illusion and Hey Rudra the lord of great blessings,

I acclaim you for your divine serenity, salutation to you O lord of endeavor forces.

Salutation to you O mother of world, the universal mother you are,

You are the universal friend helps in the sustainability of entire world.

O lord you adorn the abode of this creation; *you create, sustain and annihilate the world by your three aspects O Mahadevi salutation to you.*

You are the soul, inner soul and transcendental soul, I acclaim you.

Creation is blessed by you; universal elements are graced by you O Lord Mahadeva.

You are the coronal fire, mass ejection in the phase of Cataclysm, you annihilate the entire realm of physical world, salutation to you O Lord Mahadeva.

Impossible to know your glory; O empyrean goddess, salutation to you.

You can only recognized by transcendental soul from tough penance.

O Adi Shakti (great energy) you hold the entire cosmos by your transcendental power with different form of energies,

The entire world exist by mean of your various forms (Mass transformation) Salutation to you O Lord.

Your diverse forms of matter (Rudras), annihilates and mortify the antimatter (Devil's).

You are the strength (energy) to protect for those who come in you asylum,

O Lord you are the epitome of Fame, Bravery, Fearlessness.

World is your part and you exist apart of this world salutation to you.

I bow to Parmashwara who release and grace liberation (Moksha).

I bow to Parmashwari who grace the soul by the science of liberation.

O *Mrityunjaya Shiv*, you are the primal therapist,
who heals from deadly diseases.

You are the symbol of wisdom like a moonlight I bow to you.

Salutation to you O god of ultimate fire capable
of burning entire world,

Salutation to you O goddess Bhairvi,

Salutation to you O goddess you can ruin the three basic Gunas (qualities)

Salutation to you O primordial omniscient Lord,

Hail to you O goddess of adorning divine knowledge.

Salutation to you, you exist in everything, *Salutation to
you O goddess of desire and boon.*

Forgive me for my misdeeds.

I bow to Shive (Adi-Shakti) and Shiva, in the form of Ardhnarishvara.

Brahma ji said:- Hey Devrishi this way I have recited the glory of Ardhnarishwara the symbolism of Ekari (singularity), everything emerges from Shiv and Shive in beginning and at last everything will merge in them.

Lord Ardhnarishwara Said in a mystic voice:- Hey Prajapathi Brahaman, you really appeased me, I know your wish and yes I will do so because you have done hard penance, to *densely* populate your creation.

Thereafter Lord Ardhnarishwara divaricated into two; right is Shiva (Rudra) and left being the goddess Shive (Rudrani).

Brahma Ji Said:- Hey Narad as you already knows Shiv created numerous Rudra's as he told, He created an identical Rudras same as of him in appearance, power and behaviour, they all are immortal.

I told to Lord Shiva:- He Maha Dev, Please create the being with the cyclic order of birth and death, those all who should have fear of birth and death.

Shiva Said:- Hey Brahman I will not create the creature who has to suffer in life with all miseries of birth and death, in spite I will take out the 'Soul, Spirits and the Beings' from the miseries of sorrow and grief.

You continue to do that and create the being with your wish, because of my blessings, you will be free from any type of illusion.

Hey Brahmadev, this Devi Rudrani is capable enough to fulfil all your wishes and desires. So you request her to help.

Devrishi Narad asked:- What about Goddess Adi-Shakti Rudrani? What she has done?

Brahma Ji Said:- Hey Narad, then I started to praise numinous Rudrani, I recited to her with **Kamakshi Stuti.**

5.20 Kamakshi Stuti by Brahma

Brahma hailed Kamakshi as follows:

Jaya Devi Jaganmaatarjaya Tripurasundari, Jaya,
Shreenaatha Sahajey Jaya Shree Sarva Mangaley/

Jaya Shree Karunaa raashey Jaya Shreengaara naayikey,
Jaya jayedhika Siddheshi Jaya Yogendra vanditey/

Jayajaya Jagadamba Nitya rupey Jaya Jaya Sannuta lokasoukhya dhaatri,
Jaja Jaya Himashailakirtaneey Jaya Jaya Shankara Kaama Vaamanetri/

Jagajjanamastithi dhwamsamapidhana anugrahaanmuhuh,
yaa karoti swasankalpaattasyyai Devyai namonamah/

Varnaashramaanaam saankarya kaarinah Paapino janaan,
Nihantyaa-dyaatiteekshanaastraitasyai Devyai/

Naagamaischa na Vedaischa na Shaastrairnayogibhih,
Vedyaacha Swasamvedyaa Tasyai Devyai namonamah/

Rashasyaamnaaya Vedaantai -statwa vidbhir Muneeshwaraih,
Param Brahmeti yaa khyaataa tasyai/

Hridayasthaapi Sarveshaam yaa na kenaapi drushyatey,
Sukshma Vignaana Rupaayai/

Brahmaa Vishnuscha Rudrascha Ishwarascha Sadaa Shivah,
Yaddadhyaanaikaparaa nityamtasyai/

Yacharana bhaktaa Indraadyaa Yadaagjnaameva bibhrati,
Saamraajya sampadeeshaayai tasyai/

Vedaa nishvasitam yasyaa veekshitam Bhuta panchakam,
Smitam Charaacharam Vishwam tasyai/

Sahasra sirshaa Bhogindro Dharitreem tuyadaagnaya,
Dhattey Sarvajanaadhaaram tasyai/

Jwaladyagnistapatyarko Vaato vaatiyadaajnayaa,
Jnaana Shakti Swarupaayai tasyai/

Panca Vimshati Tatwaani Maayaa Kanchuka Panchakam,
Yanmayam Munayah praahurtasyai/

Shiva Shakti Swarupaaschaiva Shuddha bodhah Sadaa Shivah,
Yadyunmeshavibhedaah syustasyai/

Gururmantro Devataacha tathaapraanaascha panchadhaa,
Ya viraajati chidrupaa tasyai/

Sarvaatmanaamantaraatmaa Paramaananda rupiney,
Shree Vidyeyti smrutaavaa tutasyai/

Darshanaanicha sarvaani yadangaani vidurbudhaah,
Tatthaanniyama yupaayaitasyai Devyai namonamah/

Ya bhati Sarva lokeshu manimantraishadhaatmanaa,
Tatwopadesha rupaayai tasyai/

Deshakaala padaarthaatmaa yayadvastu yathaa tathaa,
Tat rupena yaa bhaati tasyai/

Hey Pratibhataakaaraa Kalyaana guna shaalini
Vishvottheerneti chaakhaataa tasyai/

Iti stutwaa Mahaa Devim Dhaata Lokapitaamahh,
Bhuyobhuyo Namaskrutya sahasaa sharanam gatah/

(Jaganmaata, Tripurasundari, Sister of Lakshmi Pati, Sarva Mangala Rupini!

You are an Embodiment of Mercy, Shreegara Nayaki, and Siddheswari, saluted by Top Yogis, Jagadamba, and Unfailing donor of boons to Devotees;

Praised by Himashaila; always visioning Shankara with desire from your left Eye, apparently as she is ArdhaNaareeswari.

My salutations to you, as you bless the deeds of Srishti-Sthiti-Layas (creation, sustainability and dissolution).

You use weapons to smash those who are responsible for violating Varnaashrama vidhis, Devi Kamakshi (Varnaashrama vidhis is the natural classifications system of all human societies to perform duties of priests, soldiers, kings, merchants, agriculture, farming etc).

You are Swatma Rupini or of form that is self- generated and neither Agamas, Vedas, Shastras, Tatwa Vettaas, nor Maha Munis. (Brahma says, you are the eternal, perpetual endless energy and a self generated impulse)

My greetings to you Devi, whom Brahma-Vishnu-Rudra-Ishwara-Sadashivas meditate and Indra and Devas prostrate.

Vedas are your 'nishvaasaas' or exhalings; Pancha Bhutas are your looks;

The Charaachara Jagat (entire world) is your smile;

It is with your instruction that AdiSesha holds the weight of Earth, Agni produces flames, Sun gives out heat and radiance; Vayu is at work to blow winds; all the Tatwas numbering twenty five are all of Devi Swarupa;

Shiva-Shakti-Ishwara-Sadashiva are all 'Unmesha Bhedas' or eye strokes;

You are Guru, Mantra, Devata and 'Pancha Praanaas', Sarvaatma, Antaratma, Paramaananda and Shree Vidya!

Sarva Darshana Shastras constitute your body parts as opined by Panditas.

You are Aayaa -niyama Swarupini; Desa-Kaala-Padartha-Vastu Rupini;

Hemamayaakaara or of Golden Form! Kalyana Guna or of Qualities of Propitiousness! You are indeed beyond the Universe!).

Hey Devrishi on my recitation Devi Adi Shakti pleased and said:

Devi Said:- Hey Brahman, I will blessed you as you desire, I will make your every word and idiom true, whatever you recited in my glory, Now tell me what else you want.

Brahma Ji Said:- Hey Omnipresent Devi, Mahadev ji has given me the responsibility of creation, I obeyed him and everytime by my mental efforts I create the creature and Devatas (Demi Deities) but there was no appropriate population mobility.

Hey Devi, graces me the power of the capability of sexual reproduction, so I can take this world into optimum population.

Hey Devi, before you there was no manifestation of feminine. Hence I am not capable of creating womanhood. All types of energy are manifested from you.

Devi, you are the one who transforms energy for various work thus I pray you to grace the power and beget as a daughter of my son Daksha.

Hey Narad, on my request, Devi with same energy and effulgence appeared from the mid of both eyebrow of Devi Rudrani. Lord Shiva smile and told to Devi:

Shiva told:- Hey Devi, you harden youself deep penance and then grace Brahmaji to accomplish his wish.

Brahmaji Said:- Hey Dev Rishi Narad, Shiva and Shive got merged again as Ardhnarishwara and **evanesced.**

Then Lord Rudra with all the Rudragana also got **evanesced** from there.

Hey Narad those all Rudras are protecting the life and creatures on various living planets throughout the cosmos.

Devi Rudrani transforms herself in various forms, pervaded throughout the region of cosmos. She performs all the tasks of mobility and action.

Lord Shiva is the Par-Brahman and Tripurasundari (Uma) is his supreme Maya (Moola Prakriti) as Parmeshwari. They are indeed one, but for the purpose of creation, they show up as two as - Uma (Prakriti) and Shiva (Purusha).

Brahma Ji said:- Hey Devrishi, "Bhagwati Rudrani" not only transform herself into Sati (She is the aspect of Adi Shakti with pride and high esteem) but also manifested as all the daughter's of Daksha after a due course of time. But before this she entered within me too.

The transformation of energy as various forms of Daksha Daughter's

Narad Ji asked:- O lord, how and why Devi entered in you, What the need of that, I want to know please let me know the entire processes?

Brahma Ji Said:- Hey Narad muni, Its highly required, to develop the Gynaecocracy, the processes of creating feminine begans from me. To create a first lady of the universe, certain matter with mass and primordial energy is required, since I am the creator hence Devi decided to start creating womanhood from me.

I went in to the state of meditation; Devi transformed her energy in me and as a consequence my physique was divided into two halves of Male and female.

From male part I was duplicated into a man and from female part women got emerged.

I named him Svayambhuva **Manu (Svyambhuva means alike me)** as he was analogous to my mind and thoughts hence called Manu (In Sanskrit Man means Mind) and he himself begot from me so called Svayambhuva.

A lady is called a **Satrupa (Sat means true and Rupa means Form or aspect)**, they are my spiritual son and daughter.

Vijay:- This is how the gender breakdown in the creation of Brahma's, all Rudras are the aspect of masculinity and Uma Devi Rudrani becomes the femininity in the worldwide.

Rudra's become the Sun, Agni, Vasus, Pusana and other Vishwa Deva's while Adi-Shakti transform herself in numerous aspects of energy like Aditi the Cosmic Matrix, Prasni the mid atmosphere, Vineeta and Kadru.

In Vedas, the deities of first generation like Indra, Agni, Savitr, Vasus etc including Sapta Rishis (Seven Seer's) are the transformation of matter in the form of Rudras, later on these responsibilities are given to the progeny's of Manas-Putras of Brahma Ji. There is nothing else in the world other than these two Rudra Shiva and Rudrani Devi Shive.

Vijay:- This is the story of origin of the life now we will evaluate and determine the significance, worth, or condition of Sarga in connection with modern science.

*I will narrate you the science mention by **Rishi Sutji** in **Garud Puran**. Once **Sonakadi Rishis**, asked Sutji about the liberation, to merge soul with Parmatama.*

Sutji Said:- Hey Sonak Rishies, Once Garuda asked to Shri Vishnu.

Garuda Said:- Life on Mrityu Loka is always miserable; no one is ever known to be happy?

O Lord of Liberation, tell me by what means they may obtain liberation?

Lord Shri Narayan said: • Listen, O Garuda, I will explain to you the mean of liberation what you have asked, even by the hearing of which a man is released* from the world of miseries and sufferings. [*By hearing alone to any discourse, liberation is not possible. What it implies is that armed with this knowledge one gets motivated to seek liberation from this ever changing world. I will also narrate you the origin of Sarga (Creation).

Reflect over what you hear and then put it in practice because liberation has to be achieved while in human body under the guidance of an adept Teacher•

*There is a Shining One, **Shiva**, who has the nature of Supreme Brahman, who is part-less, all-knowing, all-doing, Lord of all, stainless and second-less, Self- illumined, beginning-less and endless, beyond the Beyond, without attributes, Being and Knowing and Bliss. That which is considered the individual is a part of Him. • These, like sparks of a fire, with* *beginning-less ignorance, separated and encased in bodies by beginning-less karma, are fettered by forms of good and evil, giving happiness and misery, with bonds of caste, color, creed and nationality, length of life, and fortune-miseries born of karma.*

Oh Garuda, a higher and more subtle body, the Subtle-Linga-body (ethereal form of corporeal body), lasting until liberation. • Listen now the order of liberation.

Vijay:- Narayan was elaborating about the life cycle of body, Its birth, growth and death. On the other hand Subtle – Linga-body is sustaining and adopting the various species body based on the Karma of it during human form.

The Subtle-linga-body or the **Sukshma Sharira** gets liberation only after merges with **Param Shiva** or **Parbhama Narayana** as a reward by the virtue of human beings good deeds practice.

Now the science of origin of species was narrated by Shri Narayan to Garuda Ji. Let's see that.

Shri Hari Vishnu Said:- O Garud, first the unmoving things (inanimate), followed by (animate) worms, plant kingdom, birds, animals, men, the righteous, the thirty-three types deities are also the liberated, according to their order, having worn and cast aside the four sorts of bodies* thousands of times, one becomes a man by good deeds, and if he becomes a knower (one who knows the Self) he attains liberation.

Narayan further described four sorts of bodies, which are: Pindaj – a creature born with a life from the womb; Andaj – born out of an egg; Swedaj – born out of perspiration; and Udhbhij – plants that sprout from ground, vegetation.] • The embodied, in the eighty-four hundred thousands of bodies before attaining human birth, can obtain no knowledge of the truth. Through millions of myriads of thousands of births some time a being obtains human birth, through the accumulation of merit.

Vijay:- Modern Science described the nomenclature of Species with Phylum, Class, Kingdom and Genus.

Here we are discussing Brahmaji's creation which is mention in Vedas and Puranas in advance form.

Shri Hari Vishnu Ji, narrated the same science of Brahmaji creation and same was said to Devrishi Narad Ji by Prajapathi Brahma.

In today's world every one of us knows our parents, we know our grand father and mother, even some of us knows great grandfather and Great grandmother but if it is asked someone to tell his 3rd or 4th generation ancestors no one or hardly few in billions will able to answer in this world.

I am sure there is nobody in today's time that can draw their past family beyond past 5th or 6th generations.

In fact, if we do the research and draw our family tree, we will find that at last our last father and mother are Shri Manu and Devi Satrupa and they are the linage of Brahma Ji. Brahma the Cosmic Time sprouted from Lord Vishnu and Vishnu in celestial form was born from Almighty Shiv-Adi Shakti.

I mean to say here that our forefathers (Pitar) are one and they are Shiva and Shakti. This lineage in not only applicable to human being, indeed it's for the entire creature whether animate or inanimate, Vegetative, aquatic or even the cosmic constellations.

That's why we in Hinduism believe in the rituals of donating Pinda (Pind Daan) because we are blessed by the power of knowledge and intelligence because we are associated with our forefather's we wear the same genes of our ancestors.

As a mandatory rite, Pinda Daan is believed to bring salvation to departed souls; Pind-daan is must to do obligation of all Hindus or followers of Hindu religion.

With the ritual performed to pay homage to the deceased ancestors also called 'Pitar' as per our ancient texts, the religious and correct completion of this ritual helps the departed soul to attain **Moksha** (salvation).

Same way we have discussed about the celestial bodies, the entire generation of Stars are the lineage of Rudra Tar and Tara Devi.

Brahamaji said:- Hey Devrishi, before the creation of life in the form of micro organism and other living organisms, the life begins in the form of cosmic celestials, there are generation of stars, Devas, Daityas, Adityas, Nagas all living entities are linage of Rishi Kashyapa with wife Aditi, Diti, Kadru, Vineeta and so on.

While spiritual and Vedic science describe the Classification of origin of the organisms in following categories.

All the Beings or Creatures in the nature take birth through any one of the following method in materialistic world.

Brahmaji Said:- Hey Narad, When this "Brahmanda" was activated by striking the lightning force of Shri Vishnu after that I created the creature, I will narrate you the story behind this research. In the sequence of life I have created the following class of species and there are 84 (8.4 million) Lakh's form of species created by me, out of which 9 Lakh's of which are aquatic ones, 4 Lakh's are mammal species, 30 Lakh' share beasts and reptiles, 10 Lakh's are birds and 1 Cr 10 lakh's are small living insects species.

1. **Andaj:** Birth through Eggs

2. **Pindaj:** Birth through womb (stomach)

3. **Swedaj:** Birth through sweat or dirty things like some organism and micro organism. The Seedaj.

4. **Udbhij:** Birth of plants, trees or any vegetative the sprout form of life through seeds.

Apart from these four classes there are 2 more types of creatures.

Anthashatrus and **Swabhabas**, which are even subtler then virus and are the prime cause of epidemics.

These creatures are being created by Shri Brahma Ji and the delivered by different source on Mirtyu Loka.

Narad Ji asked:- O Lord, O father of world, nothing is arduous to you but is that simple or you have adopted something special to create organism.

BrahmaJi Said:- The process of creation involves various phases, as you know nothing can sustain without the blessings of vast space Shri Hari Vishnu.

I have taken the different ratio of elements but to get alive I meditate on Shri Hari Vishnu, he appeared and said.

Shri Vishnu Ji:-

Seyam devataiksata, hantaham imas tisro devata anena jivena tmana nupravisya nama-rupe vyakaravaniti.

May I enter into these three elements created by you with-fire, water and earth-and and those created by you, May I become further manifold of life, by means of the manifold of these.

Shital:- Its great to know the such a scientific classification narrated and endorsed by Shri Brahma Ji, the creation of brahma Ji, as you told is same what was endorsed by Modern Scientific theory, which are defining groups of biological organisms on the basis of shared characteristics and giving names to those groups, isn't it?

Vijay:- Yes, organisms are grouped together into taxa (singular: taxon) and these groups are given a taxonomic rank; groups of a given rank can be aggregated to form a super group of lower rank, thus creating a taxonomic hierarchy.

The same science was elaborated by modern classification of life and nomenclature under the branch of science called Taxonomy.

We will evaluate little more about these four categories of **Shri Brahmaji's** creation, namely: **Andaj, Pindaj, Swedaj and Udvhij.** These living species and it's scientific classification is mentioned in Padma Purana, the total number of living species are 84 (8.4 million) Lakh's form of species created by me, out of which 9 Lakh's of which are aquatic ones, 4 Lakh's are mammal species, 30 Lakh' share beasts and reptiles, 10 Lakh's are birds and 1 Cr 10 lakh's are small living insects species.

- **Andaj:-** The species that lay eggs (And) the beings that come out bursting through the egg are called Andaj. **The modern scientific term for an Andaj i.e.animals that lay eggs are oviparous.** The word comes from the Latin words "ova," meaning egg, and "parous," meaning bearing or producing. Egg laying is common to all species of birds, fishes, most reptiles and insects.

- **Pindaj:-** The species that give birth in the form of body (Pinda). Birth, also known as parturition, is the process in which a live, fully developed baby comes out of the mother's uterus. Mammals are the only species on earth that nourish their infants through breastfeeding. Most mammals have a placenta that feed the young ones, while they are still in the mother's uterus hence also called **viviparous animals.** Placenta, the organ that feeds the baby in the uterus from the mother's body.

- **Sweda:-** The word swedaj means the organism born by self replicating from the source of dirt and moisture like sweat. They may be visible or non visible microbes; the other Vedic word for Swedaj is *krimi* is used in *Veda* for different macroscopic & microscopic creatures. Right from bacteria, various insects like *kita, patanga* were nominated as *krimi.* The detailed classifications of Krimi's are mentioned in Vedas. Two types of *krimi* viz. *Drishta* (Visible/Macroscopic) & *Adrishta* (Invisible/ Microscopic) were described in *Vedas.* These two categories encompasses nearly all *krimi* (Microbes /pathogens). According to their origin & habitat they were categorized as *pranyashrayee & Anyasthanashrayee.* Different *sharirika, manasika* & *adhyatmic vyadhis* were thought to be originated from these *Krimis.* These harmful & debilitates (*Pushtinashaka*) organisms were recognized by various names based on troubles/sufferings they produce. Sun & *Agni* (fire) were described as internal source of *krimichikitsa.*

- Today science also confirms this fact. That early morning ultraviolet light rays emanating from sun can be used for various *krimijanya-vyadhis*. Apart from this various treatment modalities by using various natural resources; vegetable drugs, Mineral drugs, fumigation, cleansing (*Marjan-prokshana*) & hymns were described for *krimi* & diseases caused by them.

But the above mention creation of Brahmaji is not that simple, it's the creation of Brahmaji and the creation done by him involve the processes of his divine engineering of different mean. Though he has created all the Beejas (Seeds) and then later on the procedure of sexual and asexual reproduction was adopted when the life on earth was transported.

The formost Question

Shital:- You have disclosed many secret's of Vedic science and now it's clear to me about the origin of various phases of the universe from Latency Period to this vast form. You have recited the theory of Singularity, the theory of manifestation of **Brahma, Vishnu** and **Mahesha** from mass and energy to Space, Time and Matter.

Then you have narrated the Rudra and the Transcendental Power the **Adi-Shakti** in the form of **Dus Mahavidhya** (ten wisdoms). Then now you are telling about the very first beginning of life in the form of various species.

But the foremost question which is hammering my thoughts from so long was that in our society there are Sectarian believes in Shiva, some Sectarian devoted to Shri Narayan and his different swaroopa while some religious sectarianism to Adi-Shakti. Even the large folks of humans are migrated towards the spiky fanaticism of theology.

And you will agree that these fanatic societies questioned over our belief system of many deities, Can you put some rays on untold Vedic wisdom over this issue?

Vijay:- I will answer this in two ways, firstly the second point you raised about the fanatic society and their views on Hindu Dharma but I will call our religion as a **Sanatan-Dharma** the Vedic word given to the entire society in Vedic era.

Since the origin of intelligent beings, begins by the creation of Dus-Manas Putra by Brahma Ji, then Brahma Ji, created Manu and Satrupa, from Manu the linage began are called Manusyas (English world Manushya means Human). This is again the journey of humanity.

After certain period of time the society of Human being was divided into four traditional social classes (*varnas*).

The *varnas* have been known since a hymn in the Rigveda that portrays the Brahman (priest), the Kshatriya (noble or the woriers), the Vaishya (commoner or the business class), and the Shudra (the helpers).

These divisions was done when the population growth rate of early world started to rise with high speed for the purpose to fulfil own duties called Sva-Dharma and to regulate the system.

This division was done by whom and when there was a long history and story, which we will slightly discuss later on, you just remember to remained me later to start our discussion on this Vedic cult on religion.

Emphasis of Shashnag

Shital:- But it appears that Lord Vishnu is resting over the Great Serpent called Sheshnag. I want to know that there is any scientific relation of Sheshnag with the universe.

Vijay:- Everything mentioned in Vedas and Puranas are strongly connected to the universe. Sheshnag and Aditi are the vast form of Sukshm Brahm to it an extant extension.

But before our discussion begins, I want to tell you one interesting anecdote that some five years back when I was staying in hostel in Hyderabad my roommate Mr. Dev Choudhary and Vamshi asked me about the electromagnetism which controls the celestial objects.

Mr. Dev was a physics post graduate qualified student and he was doing his GATE preparation in those days.

Then I shared with them that interesting fact, which I faced during my journey from Indore to Bhopal. I was travelling with Intercity Express, in the same coach there were four passengers gossiping and was chatting on different points of Ramayana and Mahabharata.

One gentle man was a retired government first class officer while two are police-men and another one was a software student.

Their topic of discussion was randomly based on the arguments between them.

A Policeman his surname was Malviya I remember because of his batch on his dress, he said Hindu religion was one of the religions which don't have the values in their teachings.

What the nonsense is mentioned in the puranas. Men in front of him asked what do you mean?

Is it true that, Earth is located on the hood of Lord Sheshnag replied the policeman Mr. Malviya.

Retired old Man said:- Yes its possible Sheshnag is the great serpent.

What you are talking about Shasnag who knows him, it was mentioned that he keeps earth on his head. Science never accepts it. On Mr.Malviya reply his colleague nodded his head.

In between the IT student told that's why, we and our generation do not believe over the ridicules stories of Puranas.

Old Man said:- Though we worship and do rituals on regular basis, but my son is in America he also comments the same.

I was enjoying the climate and journey on my side berth, was listening all the discussion brooked between them, one another person suddenly picked the topic from Mahabharata.

So I thought at least I may help them to wash their brain by pouring some scientific knowledge because these types of controversial statement are very dangerous for our future generation.

I quit my seat and asked excuse me gentlemen, can I be the part of your discussion. You are interacting on very interesting topic with such a great knowledge.

Senior Job Retired person Said:- Oh yes welcome.

I thanked them, I seated quiet and solemn but my intensity of excitement to make them correct was very high.

Other police men told, In Mahabharata it's mentioned about Sri Krishna's Akshauhini (Sanskrit: अक्षौहिणी) –Army, how it can be in that small battle field of Kurukshtra, when it was more than present population of world.

I said, Uncle I may brief you on all the issue you all are addressing so far in Vedic-Scientific way. Your approach is based on the legacy of the semioticians.

I am happy to see at least you all are agreed with the subject, but the partial knowledge is very dangerous so let me first brief about the significance of Sheshnag as a part of your discussion.

It's a science; we need to understand the science of Vedas. Vedas are of limitless knowledge. It'ss mentioned in the Adiparva of Mahabharata in chapter 36 which, I am sharing you now, the conversation between Shri Brama Ji and Shashnag.

Brahma Ji Said:- '

"अधो महीं गच्छ भुजंगमोत्तम; सवयं तवैषा विवरं परदास्यति।
इमां धरां धारयता तवया हि मे; महत परियं शेषकृतं भविष्यति।।"

"adho maheen gachchh bhujangamottam; savayan tavaisha vivaran
paradaasyati.
imaan dharaan dhaarayata tavaya hi me; mahat pariyan sheshakrtan
bhavishyati."

O Sesha, I am exceedingly gratified with your self-denial and love of peace. But, at my command, let you do balance the earth (Planets) for the goodness of my creatures. Bear thou, O Sesha, properly and well this Earth, so unsteady with her mountains and forests, her seas and towns and retreats, so that she may be steady.'

What this Sheshnag is exactly, we need to overview before assessing anything. Sheshnag is an integralforce that keep all the four fundamental forces in balance, viz.: Gravity, Electromagnetic force, strong and weak force.

Therefore, he is holding all the planets under control of magnetic effect which is supported by gravity of Rudra Jati.

The Shash (Rest force) said to be the force which holds all the planets of the universe on his hoods constantly sing the glories of the God **Vishnu.** Yes the meaning of earth and planets on the hoods of Shashnag is because of this aspect of geomagnetism of Anant-Shash, the geomagnetism helps the tectonic plate from which the crust is formed.

Significance of fundamental forces Gravity, Electromagnetic force, Strong and Weak force are known to today's science. Sheshnag symbolize and represent the force that acts on cosmic bodies which suspending in the orbit and rotates and moves around planets within galaxies.

I hope it is clear to you, that the Vedas to be Apauruṣeya (अपौरुषेय means "not of a man", "superhuman" or "impersonal, authorless" is a term used to describe the Vedas) - Puranas are not just the story while all are rich of science.

Vijay:- Now we should come again in the main line of our discussion.

His divinity means Shankarsan, Shashnag and Ananta.

Shankarshan - Hybrid form of Shankar (Matter)

Shesh - Remainder

Ananta - Infinity

The Matter Shankarsan remained (Shesh) after Big-Crunch again attain the infinity (Ananta) as of universe by the process inflation of Shri Vishnu after Big bang.

The remainder (Shesh) which attains the infinity (Ananta) and the infinity which remains was a remainder in cyclic processes of cosmic origin and dissolution in terms of Mass (Shiva or Shankar).

The Shashnag is the force which controls all the four fundamental forces of the universe. Jathi, Mundi, Ardhmundi and Shikhandi are the forces which keep the entire world in proper manner by their forces of Gravity, Electromagnetism, Strong and Weak force.

The Shankarshan is the remainder of Lord Shankar (Shiva) who attains the infinity when the cosmos grows and he only remains at last when the cosmos shrunken is the product called Shashnag.

In relation to Lord Vishnu, Shashnag Ji came after the inflation of space and Shri Vishnu rest on the abode of the Shashnag.

5.21 The Kingdom of Adi Shakti

Shital: Oh, that's clear to me but there is little conflict in this that's why I am bewildered, If **Ananta** as Shesh is Infinity of universe, then what about **Aditi** she is also regarded as infinity of universe. Can you clarify this?

Vijay:- You are absolutely right, Today it's the beginning of Nav-Durga festival, this is the occasion of worshiping energy (Adi-Shakti) with devotion. We will discuss over the presence of primordial energy of this world.

Aditi is the phenomena of splitting of primordial energy Adi-Shakti in her vast form of cosmic matrix, which literally means the endless or the infinity.

So the Energy as infinity is Aditi while Mass as infinity is Ananta. Aditi is feminine while Ananta is masculine.

You will know the role of Aditi as we proceed further but before that if one understands the complete role of energy as lord Shiva described to "Devi Parvati" then it will become simple to know the symmetry of the universe.

Sutji has narrated the divinity of **Adi-Shakti**. How the Sources of different form of energy appeared in front of Devas in Himalayan region. When the Devas are suffered from the Demon **Tarkasur** and are in big panic.

This is the story which is going to unfold many secrets of theory of energy and the evolution of universe, the galaxies, the stars and even the life in the form of animate and inanimate objects, our actions is the intimate relation between Matter and Energy. The cosmos is filled with energy.

We have already discussed that **Adi Para Shakti** (Para means beyond everything) is the Singularities which is primordial divine energy and only exist before and after the universe along with mass and this secret was decoded by Lord Shiv Shankar to Parvati Devi in Kunjikastrotram. She is Adi-Shakti called as **AMBA** and even the mother of (trinity) Matter, Space and Time, because all these beget from Energy.

She is associated with everyone and with every object to accomplish various tasks of the world and keep the universe in operation. So the Adipara Shakti is beyond any practical, she even can't come in **perception** of wise man, deities or the scientists as a whole. Scientists from the generation of Einstein to today's modern scientists know that this cosmos came out of Singularity but they also accept that Singularity is beyond any observation.

Now we will knock some sense what Vedas says about her. Sutji is narrating the story which is having all the answers of complex question.

Initially, there are Vedas; to simplify these Vedas Shri Ved-Vyas had divided Vedas into eighteen Mahapuranas to understand the typical concept of the universe and God.

But now that causes a controversial concept by reciting these, different people started to believe in different Puranas and their supreme Deities. Like by reciting Shiv-Purana some people claims that Lord Shiva is Supreme while Some people worship Lord Vishnu having all the powers and the one who is omnipresent because there believe system is in Shrimad Bhagwat Purana.

Shri Ved-Vyas have done that work because he just wants to understand and realize the society that supreme in supremacy and that was excellent work and there is nothing wrong in that. The unfair is that when people started to spread the conflicting controversial thoughts of that this is superior to that or that is superior to this.

This Theory is of Mass, Energy, Space and then Time as we have already discussed it in detail that Supreme Deity (Lord Maha Vishnu) is operating this universe in the manifested form of these fundamentals.

Lord Maha Vishnu is the Pursha the cosmic space who operate the entire universe from his abode.

Shri Vishnu is called **Hiranyanabha**, the one from whose navel entire universe emerges.

In reality Vedas claims that lord Shiva and Vishnu are not different aspect they are one along with Adi-Shakti.

Vijay:- Today it's a great day for me.

Shital:- Oh, what happened, why it's a great day?

Vijay:- Because Shivomya is the part of today's discussion, she asked me the secret of this universe.

Shivomya Said:- Mostly I was the listener of what you narrated before to mom but today I have a curiosity about the mantra which you recite everyday i.e. Om Aim Hrim Kreem Chamunday Vichhey Namaha.

Vijay:- 'Oh Super, so I will narrate you the secret of which was once recited by Lord Shiva to Adi Shakti Parvati Devi on her request to reveal the universal code of operation.

Devi Parvati Asked:- O Lord, how this universe is in operation and how you, Shri Narayan and Brahma Ji manage this world.

Shiva Said:- Hey Devi it's you, none other than you and the significance of your aspects which operates, sustain and perish the whole universe.

Devi was surprise a while, she smiled and Said, 'O Lord' you are kidding me.

Lord Shiva:- Absolutely not, hey Devi listen the secret.

Vijay:- Lord Shiva started reciting the formula of Kunjikastrotram.

Shiva Said:-

॥सिद्धकुञ्जिकास्तोत्रम्॥

शिव उवाच

शृणु देवि प्रवक्ष्यामि, कुञ्जिकास्तोत्रमुत्तमम्।

येन मन्त्रप्रभावेण चण्डीजापः शुभो भवेत॥ १ ॥

Meaning:- Listen Devi carefully the greatness of Kunjikastrotram, It's effect is auspicious, when this Chandi-prayer is recited it blesses with luck.

न कवचं नार्गलास्तोत्रं कीलकं न रहस्यकम्।

न सूक्तं नापि ध्यानं च न न्यासो न च वार्चनम्॥ २ ॥

There is no need to recite the preliminary stotras Kavacham (for Armour), Argalam, Kilakam and the Rahasya. Not also Secrets (divulge), for this no specific method of meditation and convention is necessary.

कुञ्जिकापाठमात्रेण दुर्गापाठफलं लभेत्।

अति गुह्यतरं देवि देवानामपि दुर्लभम्॥ ३ ॥

Reciting Kunjika Stotram is enough to get the benefit of reading Durgasapta Shati. It is the most secret wonder of even the Deities don't know this mantra.

गोपनीयं प्रयत्नेन स्वयोनिरिव पार्वति।

मारणं मोहनं वश्यं स्तम्भनोच्चाटनादिकम्।

पाठमात्रेण संसिद्ध्येत् कुञ्जिकास्तोत्रमुत्तमम्॥ ४ ॥

Hey Parwati, Try (प्रयत्नेन) to keep this most secret (गोपनीयं) with you in the same way as you keep your genitalia (स्वयोनिरिव) secret, Marnam (To killing) Mohnam (illusion) Vashyam (slavery), Stambhano (paralysis by repeated chants) and Ucchatana (to send away). It is most great and auspicious by just reciting this everything will be achieved.

॥अथ मन्त्रः॥

ॐ ऐं ह्रीं क्लीं चामुण्डायै विच्चे॥

ॐ ग्लौं हुं क्लीं जूं सः ज्वालय ज्वालय ज्वल ज्वल प्रज्वल प्रज्वल
ऐं ह्रीं क्लीं चामुण्डायै विच्चे ज्वल हं सं लं क्षं फट् स्वाहा॥

AUM- the primordial sound of Almighty, AIM- Singularity (Adi-Shakti) Hreem- the energy of Protection (Maha-Lakshmi), Kilim- the energy that arouses the desire of reproduction, Chamunday- which destroyed evils (Antimatters), Vichhey- means to bless and fulfill the wishes.

Scientifically- This states the various phases of universe origin and transformation of Adi-Shakti (Energy) from singularities to the energies of different forms to create the universe. The First Energy, out of whom the entire Creation manifested and expressed herself in all diversification of nature.

It can be understood in a very simple way, which the singularity (Aim) transforms herself in the form of Hreem (the form of energy which is associated with space as prosperity) and Kreem is the energy which is responsible for the processes of desire and reproduction. This is phenomena of creation, which specify the divine mechanism of the origin of the universe.

This mantra (ॐ ग्लौं हुं क्लीं जूं सः ज्वालय ज्वालय ज्वल ज्वल प्रज्वल प्रज्वल) from Kunjikastrotram tells that how the universe evolved by the blast and scared sound AUM with the explosion of Mass (ग्लौं is the Mass, the Anunasika of Shiva and energy) with sound हुं, Kleem Adi-Shakti started to produce जूं the cosmic Space (Vishnu) with सः –that the primordial fire-ज्वालय ज्वालय with radiation - ज्वल ज्वल enlighten a glow (प्रज्वल प्रज्वल) by the energy Kleem

॥इति मन्त्रः॥

नमस्ते रूद्ररूपिण्यै नमस्ते मधुमर्दिनि।
नमः कैटभहारिण्यै नमस्ते महिषार्दिनि॥ १ ॥

Salutation to The energy in the form of Anger or ablaze (Radiation), the goddess energy associated with matter. Salutation to the energy Yoga-Maya (antibiotics as antimicrobial and antitoxins given name by modern science) that cured the most epidemic infectious infection caused by the bacteria (Kaithav) present in the puss (Madhu) cell. Yog-Maya is the energy of bonding (antigen – antibody reaction) also salutation to the energy manifested to kill the exotic Matter Mahisasur.

नमस्ते शुम्भहन्त्यै च निशुम्भासुरघातिनि।
जाग्रतं हि महादेवि जपं सिद्धं कुरूष्व मे॥ २ ॥

Salutation to the slayer of demon Shumbha and she is the one who ambushed demon Nishumbha.

O Great Goddess of divinity, awakened and blessed with grace on me, please let me become an expert of this chant.

ऐंकारी सृष्टिरूपायै ह्रींकारी प्रतिपालिका।
क्लींकारी कामरूपिण्यै बीजरूपे नमोऽस्तु ते॥ ३ ॥

Aimkari (abecedarian energy- Aim the form of energy as a creator) the originator of the universe in the form of Singularity, then the Aimkari (Singularity) transform herself in the form of energy Hreemkari (the form of the protector) to nourish the universe.

Klimkaree is the form of energy that cause desire (Different state of desire) to drive the seed of reproduction for the process of abiogenesis, salutation to that divine form.

She in Aimkari form ऐंकारी **is the singularity** who is the only one who is the only creative energy in her 1st aspect; she is the only one who created the universe.

She is the only energy in her (ह्रींकारी) form: who nurture and foster (प्रतिपालिका) the vast universe in her 2nd aspect.

In her Klimkaree (क्लींकारी) she is the mind and intellect who develop the desire to reproduce in her 3rd prime aspect.

Since, she is the energy of natural selection favours the mechanisms that result in reproduction (बीजरूपे), most significantly through the sexual or asexual urge.

चामुण्डा चण्डघाती च यैकारी वरदायिनी।
विच्चे चाभयदा नित्यं नमस्ते मन्त्ररूपिणि॥४॥

Chamunda Chandaghathi (The energy originated to destroyed the Demons) Yakari (यैकारी). The one who provide boons in her Yakari form.

Vichhay (विच्चे) who bless for every moment (नित्यं), those who chant Viche, granted protection daily. Who chants Vichhay are granted circadian protection. Salutation to her in the forms of Mantras (Codes).

धां धीं धूं धूर्जटिः पत्नी वां वीं वूं वागधीश्वरी।
क्रां क्रीं क्रूं कालिका देवि शां शीं शूं मे शुभं कुरु॥५॥

Dham, Dheem, Dhoom and Dhurjathe are the energies who are the consort of Matter (Rudra are Dhurjathas).

धूं- **Dhoomavati Devi: Dhoom Beej-Mantra of Devi Dhoomavati** is the energy in her most ablaze form, which engulf whole the cosmos within her as whole matter matter in the form of Rudra (Mahadev). Dhoomavati (Dhurjate-Patni) is the consort of Lord Shiva (Dhurjate) and Vaam, Veem, Voom, the goddess of speech and वागधीश्वरी Bagh means Loin, the energy transforms to control all the arrogant matter, she raids the Lion in her manifested form.

Kraam, Kreem **(Kali: Kreem)**, Kroom Kalika Devi is the Dark-Energy present extensively throughout the cosmos, Kraam, Kreem Kroom are the Beejmantra's of Devi Kali in her different aspects, which govern and reign the various phases of expanding universe. Saam, Sheem, Shoom, fulfill wishes along with me (Rudra-Shankar) as a Dark-Matter.

हुं हुं हुंकाररूपिण्यै जं जं जं जम्भनादिनि।
भ्रां भ्रीं भ्रूं भैरवी भद्रे भवान्यै ते नमो नमः॥६॥

Hoom, hoom Hoomkar is the Adi-Shakti who produce terrible sound energy to destroy evil anti matter. Jam jam jam jamvanadini is also the sound (Naad) the one who reproduce the auspicious elements.

Bhraam (Devi Parwati Bhramari) Bhreem, Bhroom, Bhairavi and Bhadre (Cosmic energy) are the energy of decay and illusion (Maha-Maye) known by various names.

अं कं चं टं तं पं यं शं वीं दुं ऐं वीं हं क्षं।
धिजाग्रं धिजाग्रं त्रोटय त्रोटय दीसं कुरु कुरु स्वाहा॥७॥

In Hindi and Sanskrit alphabet starts from (A) the first letter अं to last later क्षं (Aam, Kam, Tham, Tam, Pam, Yam, Sham, Veem, Dhoom, Aim, Veem, Ham, Ksham) are also the *Beej* mantras in which suffix (M) is the sound

Shiva Said:- Hey Devi, These all Beejmantras are the various forms of energy transform from Adi-Shakti that means the one and only one that is you to keep the cosmic order.

Hey Devi Parvati, tearing apart the each (words and alphabet from A to Z) Bjiamantra will sound the specific form of cosmic energy and enlighten the cosmos. you are the primordial power behind the universe is manifested from you and you hold all the forms of energies, from you only all energies are transforms in to various aspects.

पां पीं पूं पार्वती पूर्णा खां खीं खूं खेचरी तथा।
सां सीं सूं सप्तशती देव्या मन्त्रसिद्धिं कुरुष्व मे॥८॥

Hey Devi Parvati you are the daughter of the mountain, you are absolute and complete **in the aspect of** पां पीं पूं पार्वती पूर्णा - Paam, peem, poom.

Saam, Seem and Soom are the Sapthashati Beejmantra's having Mantrasiddhi's, anability to make a Mantra efficacious and to gain its desired benefit.

इदं तु कुञ्जिकास्तोत्रं मन्त्रजागर्तिहेतवे।
अभक्ते नैव दातव्यं गोपितं रक्ष पार्वति॥

So here, you yourself is the key formula of entire narration. You can fulfil the desires of devotees and you can protect them.

यस्तु कुञ्जिकाया देवि हीनां सप्तशतीं पठेत्।
न तस्य जायते सिद्धिररण्ये रोदनं यथा॥
इति श्रीरुद्रयामले गौरीतन्त्रे शिवपार्वतीसंवादे कुञ्जिकास्तोत्रं सम्पूर्णम्।
॥ॐ तत्सत्॥

By the grace of Lord Shiva and Shakti I could able to elaborate the science of Kunjikastrotram, otherwise the science of Kunjikastrotam is the knowledge beyond the human perception.

This is not just the mythological stories or the tale of human imagination, there are much more secrets in kunjikastrotram.

Shital:- "O Almighty" my senses are awakening now, You know I feel after listening the secret of Kunjikastrotram, I came out of mystic sleep and my soul has entered into the world of divine knowledge, were there is only supreme and his science.

The primordial energy Adi-Parashakti transforms herself in various forms of energies. This proof are traced by modern and ancient scientists both by which the theory of energy has been putted as energy can neither be created nor be destroyed but it can be transformed into different forms.

But can you clarify more on the line hierarchy mentioned in "Shree Siddha Kunjika Stotram" as: अं कं चं टं तं पं यं शं वीं दुं ऐं वीं हं क्षं।

Vijay:- Certainly, it's Shiv and Shiva science and which is glorifying the entire cosmos, अं कं चं टं तं पं यं शं वीं दुं ऐं वीं हं क्षं।

The Kunjika Stotram recited by Lord Shiva states that all Sanskrit Alphabets from अं **(Am)** to क्षं (Ksham) denote certain type of cosmic energy, energy within us and the energy in every aspect of the creation. The Siddha Kunjika Stotram refers to "Shree Durga Saptashati" which is the key part of "Shree Devi Mahatmyam" an abstract from Markandeya Purana. The word "Kunjika" refers to the one that has the keys to unlock the mystic secrets of the Creation. Let's deem little more.

The Siddha Kunjika Stotram is refers to Sri Durga Saptashati which the key part of Sri Devi Mahatmyam, an abstract from Markaṇḍeya Puraṇa. The word kunjika, refers to the one that has the keys to unlock the mystic secrets of the Creation

अं *(a)m* - Kamakarshini, the energy responsible to produce desire.

आ*(aa)m* - *Aditi, The Singularity which turns as an infinity.*

इ*(i)m* - Garjini, The sound energy of space at an infinity.

ई*(ee)m* - Trimurti the Devi in three aspects i.e. Sharashwati, Lakshmi, and Kali. The three forms of energies are associated with (Time Space and Matter) Brahma, Vishnu and Mahesh. She is also called a Tripursundari, can be chanted ई*(ee)m as a Beej-Mantra.*

उ*(u)m* - Vanhikavasini:- the energy of protection.

ऊ*(oo)* - Rupakarshini:- The energy of bonding, from which certain shape is formed, everything what we see is due to this form of energy.

ऋ*(r)* - Gandhakarshini:- The kinetic form of energy manifested, to which the sense organ called nose of creature react. But we do not really smell the energy; Living organism detects smells by inhaling air that contains odor molecules, for which certain energy is required.

ॠ*(r)* - Bhayankari:- This is the kinetic energy conversion from eddy available study state of potential energy, the matter react most intense in the vicinity and taken some time the most terrible forms. There are many examples when sudden storms broke out, the earth trembles, an event causing great and often sudden natural distress.

ऌ*(lr)* - Chittakarshini and Sanharini:- The energy by which mind or intellect reacts to the body.

The **brain** of every being on this world contains billions of nerve cells and sensory organs arranged in pattern that coordinate thought emotion, behavior, movement and sensation. A complicated highway system of nerves connects **brain** to the rest of the body, so communication can occur in a fraction of a second.

In the case of human, brain is made up of approximately 100 billion nerve cells, called neurons. **Neurons** have the amazing ability to gather and transmit electric and chemical signals.

So these electrochemical systems are generated by energy called लृ*(lr)* – Chittakarshini which keep mind and body connected.

Lruum – Dhairyakarshini, Karalini and Kamla:- this form of Devi energy is harmonizing essential mindsets and courage required for various activities.

ए*(e)* - Ekadashi, Vanhi and Udhvarkeshi

ऐ*(ai)* - Sarasvati, Vijaya, Dvadashi, Ugrabhairavi, Yoni, Veda

ओ*(o)* - Beejakarshini, the energy associated with seed growth.

Estrogen and testosterone tend to take centre-stage as pheromones offer a subconscious whiff of fertility resulting in lustful attraction. Sexual chemistry, however, is further developed as dopamine, norepinephrine, and serotonin promote attraction, and oxytocin and vasopressin promote bonding with a partner.

औ*(au)* - Jvalini, Atmakarshini, Dakini

अं (am) - Amrutakarshini

अ: (aha) - Chandika and Rath

क *(ka)* - Mahakali, Skandha, Kameshvari, Krodhish

ख *(kha)* - Akash, Tapini (Sound Energy and Radiant energy)

ग *(ga)* - Ganga, Vishvamata, Bhogini

घ *(gha)* - Trailokyavidya

ङ*(nga)* - *The energy of wisdom also associated with holy river Ganga.*

च*(cha)* - Vadhu, Chandrama, Kulavati, Jvalamukhi

छ*(chha)* - Sadashiv, Vilasini, Raktadanshtra

ज*(ja)* - Nandi, Bhogada, Vijaya

झ*(jha)* - Gruha, Dravini (The energy is transferred as a result of work done by gravity or Jati.

ञ*(gya)* - Vidyunmukha (Electrical energy is stored in charged particles inside an electric field.)

ट*(ta)* - Pruthvi, Marut (Marut is the Wind energy)

ठ*(tha)* - Vanhi, Kapali (Solar Energy)

ड(dham) - Bhivakra, Yogini, Bhishana (Bond Energy)

ढ(dha) - Malini, Dharni, balancing the earth and earth like planets.

ण*(na)* - *Energy involves in the expansion of ether and space.*

त*(tam)* - This Beej-Mantra is for getting rid of disease, worry, fear and illusion.

थ*(tham)* - Beej mantra in the worship of the greatly alluring Chandika.

द*(dam)* - Shankhini

ध*(dham)* - The Beej mantra associated with Yogni.

न*(na)* - Jwalini, Sinhanadi (Heat Energy)

प(pam) - Kalaratri (Dark Energy)

फ(pham) - Pralayagni, Kalakubjini (Nuclear Energy which can destroy whole cosmic structure).

ब(bam) - Kledini, Tapini, Bhayankara (Electrical energy)

भ(bham) - Bahurupi (Potential energy which can transform in various forms of energies)

म(mam) - Kali, Matangamalini

य(ya) - Sthiratma (Static Energy)

र(ra) - Agni, Krodhini, Tripursundari (Thermal energy)

ल(la) - Amruta, Pruthvi (These are Magnetic energies)

व(va) - **Seed sound:** VAM-when chanted in the presence of water, it enhances the production and circulation of fluids in the body.

श(sha) ष(sha) स(sh) ह(ha)

This way every letter of Sanskrit is the Beej-Mantra which helps us to reminiscence Goddess Adi Shakti Durga. The 51 matrikas (letters of the Sanskrit alphabet) constitute the Deity in her feminal in the form of sound.

Shivomya:- I **know** that every **good** and **excellent** thing in the world are because of energy within us, it's because of moment by moment observation, by feelings and by realising the things and that is because of Adi-Shakti and her potential.

Shital:- Every pace of the entire cosmos have certain energy and energy of a system can be subdivided and classified into potential energy, kinetic energy, or combinations of the two in various ways. Kinetic energy is determined by the movement of an object – or the composite motion of the components of an object – and potential energy reflects the potential of an object to have motion and generally is a function of the position of an object within a field or may be stored in the field itself.

The various forms of energies mentioned in the Beej-Mantras of Kunjikastotram are Kinetic energies of Adi Shakti's aspect of potential energy.

Vijay:- Energy can exist in many different forms. All forms of energy are either kinetic or potential. The energy associated with motion is called kinetic energy. The energy associated with position is called potential energy. Potential energy is the energy by virtue of an object's position relative to associated objects. Energy can be stored in motion just as it can be stored in position.

* * *

Lord Shiva Said, Hey Devi Parvathi, you are पां पीं पूं पार्वती पूर्णा. This means Devi parvati have all potential (Purna means Total sum of all

Potential energy, is often associated with restoring forces such as the force of gravity (Jati)). She can transform to various forms of kinetic energies and associated with all animate and inanimate objects.

* * *

Vijay:- You can note that Kinetic and Potential two important forms of energy are associated with the daily needs.

- o **Kinetic energy – motion** mechanical energy — motion of macroscopic systems

    - Machines

    - Wind energy

    - Wave energy

    - Sound (sonic, acoustic) energy

- o Thermal energy — motion of particles of matter

    - Geothermal energy

- o Electrical energy — motion of charges

    - Household current

    - Lightning

- o Electromagnetic radiation — Disturbance of electric and magnetic fields (classical physics) or the motion of photons (quantum physics)

    - Radio, Microwaves, Infrared, Light, Ultraviolet, X-rays, Gamma rays

    - Solar energy

Potential energy (Adi Shakti Parvati in the form of potential energies)

- Potential energy — position

- Gravitational potential energy

- ○ Roller coaster
- ○ Waterwheel
- ○ Hydroelectric power
- Electromagnetic potential energy
  - ○ Electric potential energy
  - ○ Magnetic potential energy
  - ○ Chemical potential energy
  - ○ Elastic potential energy
- Strong nuclear potential energy
  - ○ Nuclear power
  - ○ Nuclear weapons
- Weak nuclear potential energy
  - ○ Radioactive decay

Shital:- Firmly agreed, the puzzled curiosity on the features of Adi Shakti made Shivomya desired and you reveal the science of wonders.

Vijay:- Why I reveal Kunjikastotram at this juncture of our discussion just to understand sequence of timeline of the universe in better way.

Still there are lots of hidden secrets in each alphabet of Kunjikastotram, which even I am unable to express. But now I will reveal one of the most amazing facts of potentially energised Adi Shakti **Bhuvaneshwari** in her furious form.

Shital:- Furious aspect of Adi Shakti, you already narrated the aspect of **Chhinnamasta** Devi, She was also shown in her furious form, later on she was calm and pacified?

Vijay:- Off course, I have narrated somewhere a couple of months back to my friend **Sachin Soni**, it was the day we are roaming in Udaipur city, Rajasthan. He asked about the history of Galaxies.

We discussed about various papers published on this. It was night around 9:30 at the top of hill in a dark night I was helping him in spotting **Andromeda galaxy.**

Shital:- Oh, that's great to know, how people are eagerly waiting you to decode the astronomy in your own way, can you please reveal the same to me?

Vijay:- This is in continuation to the timeline of early universe, the study of the *early universe* is one of the most exciting fields of science, aided by advances in computer technology and observations from space, astronomers have begun to unravel the mysteries of galaxy formation and evolution. Galaxies evolves by interacting with their environment and especially with each other.

Fierce galactic confronts, gravitational forces generate strong tides that survive as telltale marks for billions of years. Because these so-called collisions dissipate orbital energy, galaxies on bound orbits may eventually merge. Collisions and mergers are liable for a great variety of phenomena, including the eliciting of widespread star formation in galaxies and the fuelling of nuclear activity in quasars.

Evidence shows that not all galaxies formed shortly after the Big Bang. A sizable fraction of them may have formed later and many are still experiencing significant dynamical evolution. Astronomers may observe all this by the help of computer technology but literally all these evidences are coded in ancient Indian Vedas and Puranas. Countries like Russia rich in cosmology firmly believe on Vedas and are working on the hidden science of Vedas.

Shital:- So there is significant elaborated code for the evolution of galactic encounters in Vedic phenomena and as you told that there is separation of Mass and Energy on the urge of Shri Brahama Ji. I want to know, whether Shiv and Shiva get united again and what are the consequences of their separation?

Chapter 6

Devi Dhumavati – The Collision of Galaxies

Shital:- Well stated, practically mass and energy can never be separated, But the mass defect is exactly related to the change in energy by Einstein's equation: $E = mc2$

I am interested to know what exactly happened in the early history of the universe and what are the effects of mass and energy defect?

Vijay:- It's about the Shiv and Shakti's Ardhnarishwara Swaroopa and just by knowing about severance, their separation and parting as a distinct personality, which cause the Adi Shakti to adopt the aspect of Dhumavati.

I will elaborate the science of Megamerger which seems to be a separation of Adi Shakti and Shiva in the form of Galaxies, but actually it's not a separation in fact it's a megamerger of Mass and Energy in the form of Adi Shakti and Shiv.

Shital, if you remember we have discussed the **Bhuvaneshwari** Devi's 4th aspect among Dus Mahavidhiyas.

Shital:- Bhuvaneshwari Devi is the aspect of Adi- Shakti responsible for the growth, evolution and adaptation of fourteen Galaxies.

Vijay:- Can you reveal the terse, the summary of Maha Vidhyas we discussed up to now?

Shital:- Exactly, Started from **Kali** the phase of Dark Age, transformation of Kali and the appearance of 1st Star of the Universe in the form of **Tara Devi and the universe become transperant,** From her, the origin of first generation of stars, made up of **only hydrogen and helium, these so-called population III stars,** which is precisely

3rd Maha Vidhya in the form of **Tripura Sundary** and then the ages of evolution of Galaxies begins and the formation of Fourteen Galaxies one after another, collectively called as Devi Bhuvaneshwari.

Vijay:- Certainly, These fourteen Galaxies exposed a giant collision, do you know Scientists spot huge, ancient collision in space that could change our understanding of the universe.

Shital:- But How this collision is related to Devi Bhuvaneshwari and Devi Dhumavati is the mistress of Galaxies, moreover in your earlier discussion the Galaxies are continuously receding from each other. My question is that when the galaxies are moving far from each other thus what cause such a vast collision.

Vijay:- In a real sense, Hubble's law, the recession velocity of galaxies, is an illusion. The galaxies are not moving, the space between them is literally expanded. To see how this produces a Doppler effect, consider a simply Universe that is a circle. To the observers in this type of Universe, they believe they live in a 1D structure. But, in fact, they live in a 2D structure, a circle. The position of the galaxies can be measured by the distance between them (see diagram below) or what are called the co-moving coordinates, an angle θ between the galaxies.

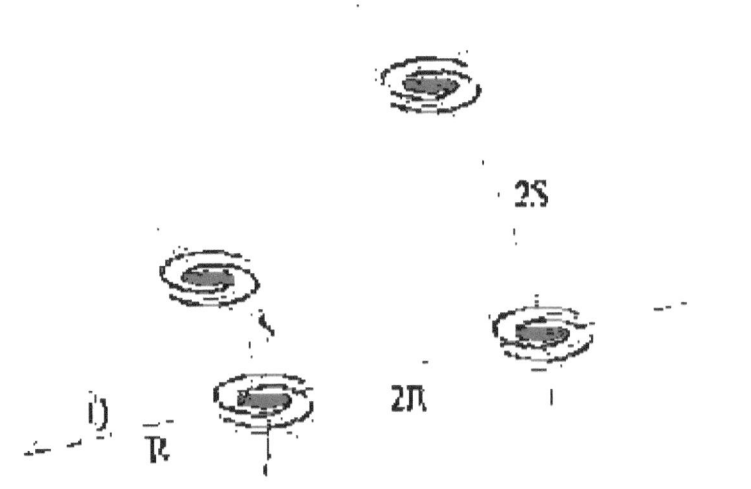

The radius of the Universe is given by R, notice that R is a quantity only seen in 2D space, not measured directly by the inhabitants of the 1D circle unless they measure $2\pi R$ by walking around the Universe. Now, we let the Universe expand by a factor of 2, R becomes 2R. The distance between the galaxies becomes 2S, but the co-moving coordinate, angle θ remains unchanged. Since the distance between the galaxies has increased, then the galaxies will appear to have moved apart by S/time of expansion. When, in fact, the galaxies have not moved at all, the space between them has increased.

Reference: Hubble's law

Vijay:-Do you know the event which is coded in *Devi Bhagvata Puran* is now marked and first time scientists observed the birth of a galaxy cluster, with at least 14 galaxies crammed into an area only about four times the size of our average-sized Milky Way galaxy.

The same events are rooted evidently in *Mahabhagvata Purana*. **This Megamerger of 14 Galaxies could become The Most Massive Structure in Our Universe and this was executed by the energy and power of Energy Dhumavati.**

But as her attribute says, Dhumavati is the primordial darkness and ignorance, from which rises the world of illusion. She represents the darkness/ignorance before creation and after decay. This ignorance, which obscures the ultimate reality, is necessary because without the realization of this ignorance, indeed true knowledge cannot be attained.

Dhumavati also represents yogic sleep (Yoganidra), the pre-creation state of consciousness, as well as the primal sleep (the Void) in which all creation would dissolve and clinches ultimate reality of *Brahman*. This void is pure consciousness, the cessation of fluctuation of the mind, and silence.

Dhumavati is often said to appear at the time of *Maha-pralaya*, the great dissolution of the cosmos and is equated with the dark clouds that rise during *Pralaya*. Her thousand name hymn also signifies her by names meaning "she who's Form is Pralaya". "She is indulges with Pralaya

Dhumavati represents ultimate destruction, literally "the smoky one", she manifest herself at the time of cosmic cataclysm.

Vijay:- I have gone through the many scripts of current curriculum in these, Devi Dhumavati is attributed as the Numinous of disappointment (she is commonly known as the deity of disappointment).

Shital:- Isn't it?

Vijay:- Of course not, indeed she is the deity of recreation; she is the wisdom of looking beyond the present. तंत्र ग्रंथो के अनुसार धूमावती ही उग्रतारा हैं. In Vedic and Puranic cult, She is worshiped for ultimate realty. The modern Astronomy described this incident with the reference as:

26 APR 2018

Peering deep into space — an astounding 90% of the way across the observable Universe — two groups of astronomers led by University of Edinburgh's Dr. Iván Oteo and Tim Miller **from Dalhousie and Yale Universities have witnessed the beginnings of gargantuan cosmic pileups: the impending collisions of young, star-forming galaxies.** Using some of the most powerful telescopes in operation today, an international research team discovered the extremely dense concentration of hot galaxies shooting towards each other.

Peering billions of light-years back to when the Universe was just 10 percent of its current age, astronomers have spotted a colossal pile-up: 14 young star bursting galaxies merging into one of the most massive structures in the Universe.

Shital:- So, What cause these galaxies mega merger, is it Devi **Bhuvaneshwari?**

Vijay:- Devi **Bhuvaneshwari** is the Divine Mother as the Queen of all Worlds, The entire Universe is said to be her pretty creation and all beings are ornaments of her infinite being. She carries entire worlds as ornaments of her own Self-nature.

Again when Shri Brahma Ji worried about an imbalance of cosmos, He penance to lord Shiva, but the lord Shiva is in deep meditation, therefore Brahma Ji praise to Adi-Shakti Devi **Bhuvaneshwari**, Devi appeared and saw that Brahma ji is worried and in trouble.

Devi Bhuvaneshwari Said:- Hey Brahman I know the reason of your distress, Tell me what you want.

Brahma Ji Said:- Hey Devi, The fourteen galaxies (Bhuwan's) created by you are devoid of appropriate growth and not fit for my creation for life and Lord Shiv Shankar Rudra is in the state of meditation. His present state might be because of your separation from him. It's my fault, I am responsible for this, Devi I apologies.

Devi Said:- Firstly, its worldly illusion, I and Lord Shiva was never separated, Mass and Energy are the two aspect of supreme, we are one and will always remain one.

Secondly, Shiva is all encompassing the universal soul and consciousness, he is lord of world (Rudra Shankar) is the one who can expedite the celestial growth.

Brahma Ji Said:- Hey Devrishi, By listening this, Devi **Bhuvaneshwari** said, Hey Brahman as you said all fourteen galaxies are not perfect fit and good enough for you to nurture the processes of creation. I will show you how Lord Shiva and myself is always united and I will vanish all these galaxies as you rightly said all fourteen galaxies are not appropriate for you, therefore the objects which won't have any use should not required.

Thereafter Devi **Bhuvaneshwari** become anguish and shown the vast and furious swaroopa (form) and her mouth is like blackhole swallowed lord Shiva in the same state of his meditative.

Shital:- Oh no, is it possible to devour Lord Shiva?

Vijay:- This is what I want to narrate when Lord Brahma won't understand pageant of Shiv-Shakti, how can we an ordinary human. Take a look on it.

Brahama Ji Said:- Hey Narad, by this abrupt episode of Adi Shakti the syncretised cosmos was seems to be collapsed. While lord Shiva was devoured, his all four form of **Jati, Mundi, Shikhandi** and **Ardhamundi** started furling with Ishan Shiva within Adi Shakti.

Since Aghore is also Shiva's another aspect, his four son's Krishna, **Krishnashik, Krishnasya and Krishnakantdhak** started pairing with basic radical axis **Lord Shiv Sankara Aghora.**

I am not telling this based on any ideology, precisely it is mentioned in Devi Bhagwat Purana.

This event was spotted by astronomers of our age says, eventually the megamerger formed a cluster of galaxies by the gravitationally bound of dark matter and ultimately smooching together into one ginormous galaxy.

Certainly by the collision of all fourteen galaxies leads to a megamerger and Lord Brahma ji was flustered by witnessing the quaint scene of Adi Shakti **Bhuvaneshwari** and Adi Deva Mahadeva, therefore Brahma Ji started recalling and memorizing Lord Narayan.

Lord Narayan appeared and appeases Shri Bhahma Ji and Said.

Lord Narayan Said:- Hey Brahaman, I know the reason why you recalled me. You are so tensed at this moment; tell me what I can do for you?

Shri Brahma Ji Said:- Hey Narrayan, the world is almost come to an end. I praise Devi **Bhuvaneshwari**, I saw her great reverence and adoration, but instead of her grace, I am facing her wrath and she devoured Lord Rudra in rage. Oh my Lord now you are only hopes for me.

Lord Narayan smile and Said:- Hey Brahaman, Who knows the pageant of Shiv and Shakti, and look at there all fourteen galaxies are collapsing.

Vijay:- Meanwhile, all fourteen galaxies got already collapsed and started forming a cluster of galaxies by the strong gravitational pull of Rudra Jati which bound the Rudra Aghora as a Dark Matter.

Eventually the megamerger formed a cluster of galaxies by the gravitationally bound of dark matter (Aghora) and ultimately smooshing together into one ginormous galaxy. This was reported in the journal published 25th Apr'18 in phys.org, on 26 APR 2018 in science alert and media coverage by breakingnews.org,

This stage of the merger is called a protocluster, and it's an extraordinary discovery.

"Having caught a massive galaxy cluster in throe of formation is spectacular in and of itself," said Scott Chapman, an astrophysicist at Dalhousie University, published in *Nature*.

"But the fact that this is happening so early in the history of the Universe poses a formidable challenge to our present-day understanding of the way structures form in the universe."

All 14 galaxies in SPT2349-56 imaged by ALMA. (ALMA (ESO/NAOJ/NRAO); B. Saxton (NRAO/AUI/NSF)

Collapsed 14 galaxies detected by ALMA (Atacama large
Millimeter/submillimeter Array) telescopes

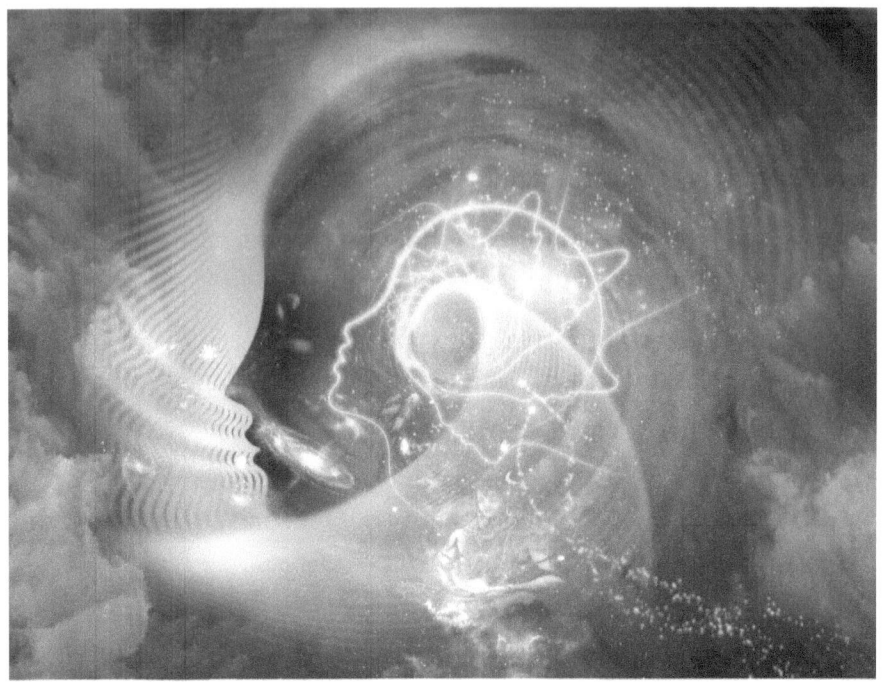

Collision of 14 Galaxies (Choudha Bhuwan) by Devi-Dhumavati)

Images designed by Bharat Bhushan

The protocluster are named as *Choudha Bhuban* (*चौदह भुवन), and astronomically named SPT2349-56, The astronomer realize that these galaxies are violently collide and merge each other is 12.4 billion light-years away, populated by dusty galaxies that are forming stars at a furious rate - up to 1,000 times faster than the Milky Way. Yet they're crammed into a space just three times bigger than our whole galaxy.

This is not a natural phenomenon, therefore the whole processes is commenced by Energy Devi Dhumavati transformed from energy **Bhuvaneshwari**.

In itself, the protocluster would be a rare find, you can fully expect to discover all kinds of things forming in the early Universe - stars, galaxies, clusters of galaxies - but the size and composition of these protoclusters is a conundrum.

"The lifetime of dusty starbursts is thought to be relatively short, because they consume their gas at an extraordinary rate," explained astrophysicist Iván Oteo from the University of Edinburgh, lead author on the arXiv paper.

"At any time, in any corner of the universe, these galaxies are usually in the minority. So, finding numerous dusty starbursts shining at the same time is very puzzling and something that we still need to understand."

After the Big Bang, according to our current models of the Universe, everything was still dark for a while. It wasn't until around 1 billion years later that the Universe became fully ionised and transparent, and we see the first galaxies start appearing.

These clusters appeared about 1.4 billion years after the Big Bang. The models of the Universe's evolution predict that, while these clusters can exist, they ought to have taken much longer than that to evolve.

"How this assembly of galaxies got so big so fast is a mystery," said Tim Miller, a PhD candidate at Yale University, and lead author on the *Nature* paper.

"It wasn't built up gradually over billions of years, as astronomers might expect. This discovery provides a great opportunity to study how massive galaxies came together to build enormous galaxy clusters."

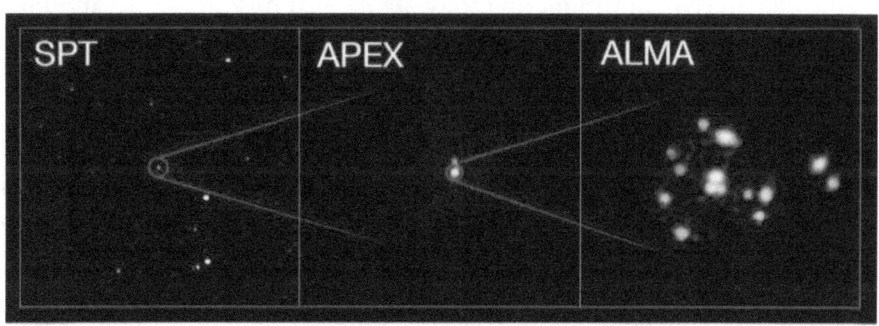

(ESO/ALMA (ESO/NAOJ/NRAO)/Miller et al.)

SPT2349-56 was first seen as a faint smudge of light imaged by the South Pole Telescope in 2010, but it was unusual enough to warrant further investigation with something more powerful.

The European Southern Observatory's (ESO) Atacama Large Millimetre Array (ALMA) and the Atacama Pathfinder Experiment (APEX) telescopes were then used to image the object in higher resolution, showing more details.

Often objects that early in the Universe are too faint for our telescopes to pick up, but there may be more of these protoclusters out there, the researchers said.

"These discoveries by ALMA are only the tip of the iceberg. Additional observations with the APEX telescope show that the real number of star-forming galaxies is likely even three times higher," said ESO astronomer Carlos De Breuck.

"Ongoing observations with the MUSE instrument on ESO's VLT are also identifying additional galaxies."

The SPT2349-56 paper has been published in the journal *Nature*,

Shital:- So what was the ultimate fate of those 14 galaxies and is Devi Dhumavati pacified?

Vijay:- But there is another twist to the story of this scientific phenomena. We have discussed seven cosmic powers out of ten Maha Vidhyas (the ten cosmic powers), some of these great cosmic energies seems distructive, but in factual they all are serenity in nature and play major role in the chronology of the universe.

The universe is the essence of all ten great cosmic powers and the universal forces of Rudra's. The Great Goddess, the power who creates and destroys everything, often represented as the mysterious metra and matrix into which everything dissolves and from which everything emerges.

To know the consequence of this megamerger, you and the audience have to wait for some time where we will discuss about the remaining three Maha-Vidhyas and major epochs are delineated by Rudras and the incarnation of Shri Hari Vishnu, each corresponding to a major period in the history of the Universe. There you will find an interesting asymmetry war between the matter and antimatter.